Nature, Science and Modern Life

Nature, Science and Modern Life

Dr. Y.S. Rajan

DIVE IN THE OCEAN OF KNOWLEDGE
Ocean Books

Ocean Books Pvt. Ltd.
ISO 9001:2015 Publishers

Published by
Ocean Books (P) Ltd.
4/19 Asaf Ali Road,
New Delhi-110 002 (INDIA)
e-mail: info@oceanbooks.in

ISBN 978-81-8430-624-8
NATURE, SCIENCE AND MODERN LIFE
by Dr. Y.S. Rajan

Edition
2025

Price
₹ 600.00 (Rupees Six Hundred only)

Printed at
Narula Printers, Delhi

Foreword

Most of us today in 21st Century love to go back to NATURE. This desire is built into our genes through evolution and wired into our brains. We no doubt are tied to our modern gadgets. But we love GREEN plants, leaves, lawns, and colours of FLOWERS. We are thrilled by the roaring oceans in the sea-coast. We love gushing waters. We love our pets. Of course we love to see the mountains and are awe-inspired by their immensity. Very rarely we get a chance to see the star-studded sky and are thrilled when we get a view of it. We fall in love with FULL MOON and are afraid of TOTAL DARKNESS!

This love for NATURE is irrespective of the fact that we occasionally come across the terrible face of NATURE: Floods, earthquakes, landslides etc. After sometime we forget it.

This love for NATURE is also partly because of the stresses and strains of modern social and economic life. We are running around all the while, despite traffic jams and long distances of travel. Many of us don't get even good sleep. So many things to engage us: office or work; TV; social media; the mobile talk, gossip; seminar talks; dead- lines to meet...No time to relax!!

So we love to go to "NATURE" for a few days in a year during our vacations. For regular daily life we grab anything with "NATURAL" title provided the price is affordable! Natural food; Natural soap; Natural health care; etc.

But herein lay some dangers too. There are some persons (through ignorance or for deliberate exploitation) who use this desire to sell many products or "ideas" which are harmful to the individuals and life in general. Some of them blame everything on modern science, technology, engineering or medical practices.

"GO BACK TO NATURE!" "GREEN Revolution was a danger; go to age old agriculture!".

"Avoid all modern medicines and vaccinations!". "Avoid GM crops", "No to plastics", etc.

We tend to forget that human population has grown several HUNDRED fold since those "good old days"!! In fact post independence, that is from mid 1940's, developing countries including India have increased their populations FIVE times; five hundred percent!

So the solutions for the modern human beings are very different from the "good old days"! They require a good understanding of Science, Engineering, Biology, Neurosciences, Economics, etc.

Yes we have to live with nature, but in the modern world which will keep changing!

The opportunity to share my ideas on this crucial theme came through the kind request to me by the (late) Arulchelvar Dr. N. Mahalingam to write in Kisan World (monthly). Each article is well researched in topics ranging from soil biology to garbage to modern materials to energy to e-commerce to geopolitics. I learnt a lot. I wrote for five years.

I thank also Smt. Shailaja Krishnamurthy of ISROHQ who meticulously typed the written manuscripts and sent to Kisan World in time all through five years. I thank my wife Gomati, sons, daughter-in-law and grandsons for their encouragement.

I request young, middle aged and senior persons who read this book to absorb what is given in these chapters and apply their minds FURTHER to shape the new forms for adapting to modern life. Entrepreneurs can create many new Start-ups which can create products and services suited to modern life and are closer to nature. They may win global markets as well.

Write to me your ideas. Let us be in touch. I will learn more!!! My email is ysrajan1944@gmail.com.

Also visit my website: www.ysrajan.com

—Dr. Y.S. Rajan

1st Jan. 2019

Contents

1

Youthful Indians: What is Future?

When I go on lecture tours to many parts of India (schools, colleges and public places) I generally use an 'interaction method' instead of monologue lectures, followed by a "formality sake questions and answers" (mostly prearranged!!). Such an interaction is richly rewarding to a speaker and the audience. It is not pre-rehearsed nor is it pre-orchestrated. I get many new insights about what the audience is thinking.

Since they have to be spontaneous in answering my questions, which are basically linked to the topic of the talk, their perceptions come out truthfully. What they say, may be right or wrong, but that is the way they feel, due to a number of reasons. I use this method right from school children, college students, to adults of various age groups.

Faith in Future of India

Two questions, I put to the audience often are:

(i) How many of you have faith in the future of India? Raise hands, if YES.

After that is over:

(ii) Let India be what it is, how many of you have faith in your own future? Raise hands, if YES.

Whether it is an audience of 100 to 200 or 1000 to 2000, the percentage of persons raising hands for question (i) has been always less than 10% almost all over India, irrespective of age group. For question (ii) it has been around 20 to 25%, all over India sometimes 30%.

I prod others who would about 65 to 70% of the audience, say two-thirds, to take a position. They continue to be non-committal.

I continue my talk trying to explain to them as to why perhaps they feel so. Some interactions, then I go on to explain what ought to be done etc.,

The main point, I wish to bring out here is that most Indians in the lower, middle and upper middle classes who are the privileged 30% of the total population, (as I speak in secondary schools and higher institutions) have very little faith in the 'Future of India.' Two thirds are non-committal. About a quarter to third feel that they can have a good future, whatever be the future of India.

What happens to the two thirds who keep quiet? Not sure about the future but ready to face whatever comes or unfolds in the months, years ahead of them?

I have often wondered what would be the answer to my queries, if I address these questions to the **real bulk of India between the ages 13 to 35 years** (they are the underprivileged bulky base ranging from 70% of youth) – all of them, not just in schools, and colleges be it in English, Tamil or Hindi medium (I speak to such students); they are below the lower middle classes in income levels. **In this article I explore their situation.**

Persons between 13 to 35 years (defined in Government of India statistics as youth) are those who are born in 1978 to 2000. They form about 40% of the total population of India. One can safely state that about 70% of them would not have had beyond 5 years of schooling or definitely not beyond the elementary school (even with RTE).

For those 70% of the youth in the Bulky Base in the age band 13 to 35 years, that is, about 350 million of Indians, what is the hope for future? They have not even been trained to acquire the skills needed for the modern economy (not their fault), which was (is) emerging in India, due to globalisation and liberalisation and worldwide technological advances. Boys when they reach 13 years of age rush to nearby towns and cities and also metropolis in search of some jobs, to get their incomes on a daily basis. If they had passed a Board exam, they invariably get out of the village. Girls of that age group (13 to 15 years) do not have that freedom. They have to stay back home in the village till they marry. Post-marriage, many of them move to urban slums with their husbands. There is a step up in their incomes, freedoms and aspirations. This is true of most of the villages as many of them are still struggling under the subsistence economy of the past. I state these based on informal talks in many villages in India in Tamil Nadu, UP, Bihar, Punjab etc., You will later find in this article that it is supported by researched articles too.

Since the investments in agriculture and related infrastructure have been inhibited due to many wrong and archaic government policies, economic growth in villages is insufficient to support the youth. Under the wrong notion that villages should not be industrialised, there had been (is) no structured programmes to have village based industries - Micro, Small and Medium Enterprises (MSME's) which can gainfully employ youth (both men and women). Only productive work can support better wages for many persons, thus creating a market for goods in rural areas. Of course, even now rural consumption of goods like TVs, cycles, beverages, soaps, shampoos etc., is growing; but, part of the money for such consumption is not through money earned in villages. It is through the "money order economy". Those who go to towns, even staying in slums, send some savings back to villages. It is good also. But bulk of the income generating jobs is in towns, cities and metropolis. As we will see in the statistics, (I will quote a little bit in this article), such availability in urban growth centres is grossly inadequate. Solution is not by pushing them back to agriculture. You will see from the statistics, why so?

But one of the most critical problems facing India now, is the joblessness of this youth. Unfortunately it is true in 2016 too.

Some Definitions and Sources

The statistical and analytical conclusions for this article are derived from the following:

(i) Government of India, Ministry of Home Affairs, information available in the Internet on Census data and their specialised interpretations and tables. They give plenty of state wise information too and also comparisons over years 1990's till now. They run to several thousands of pages. I have made selective reading. One of it is Provisional Population Totals – India -2011- a chapter State of Literacy; contain data and analysis 1901 to 2011.

(ii) An excellent report on Youth Employment and Unemployment in India by S. Mahendra Dev and M. Venkatanarayana brought out by Indira Gandhi Institute of Development Research, Mumbai April 2011 (WP-2011-009) http://www.igidr.ac.in/pdf/publication/wp-2011-009.pdp. I will refer it as SMD/MV report in this article, abbreviating the author's names.

(iii) Skilling India: The Billion People Challenge by CRISIL, November 2010. (https://www.scribd.com/document/255995839/Billion-People-Challenge).

I have not referred to several other studies and reports as well as my own articles/books which have shaped my ideas. My role has been to synthesise thoughts and provide an integrative picture and suggest solutions to this grave problem (danger) to India.

Definition of Youth: The United Nations (UN) documents, International Labour Organisation (ILO) and such bodies define youth to be those between 15 and 24, a transitional period from childhood to adulthood. Government of India (GOI) officially defines youth as between 13 and 35 years of age; but some other GOI documents like National Youth Policy of India define as age around 10 to 34 years (see more in SMD/MV report). We will use both definitions to consider actions and focus on issues.

Definition of Labour Force: Persons in the labour force are those who are **willing to work and are available** for the labour market. By this definition all those who are jobless and unemployed also form the **Labour Force.** This is Potential Human Resource.

When studying the statistics with the above definition, one will find (as I was during late 1990's when I was conceptualising my book Choosing Career Paths) that the Labour force does not increase at the rate of increase of population in the entry age group whether one takes 15 years (which is the normal year of economic statistics or 18 years etc.,) some of this can be attributed to the fact that those who reach 15 years or 18 years of age, **may not be willing to work because of continuation of their studies.** After discounting this number also, which is a small percentage as the bulk of India's youth never cross beyond matriculation (10th Board), I found the numbers are not tallying with the increase in India's youth population which is approximately **20 million each year.**

Whatever the entry cut off age one puts 13 years, 15 years or 18 years; I found out later that this is because of **exclusion of women** as a (potential) labour force using the past trends. Girls/young women are barred from going to work due to various restrictions: community taboos, restrictions by certain religious groups, their own fears as their mothers and grandmothers have not gone for work, demands from family etc.,

Working women who take long remission due to pregnancy/ child care still will appear as Labour Force though shown as unemployed.

I am addressing this **Women Skew** in this article and am giving an approximate correction factor. None of the above quoted references have done so, as they have kept the traditional continuity. But I have derived the correction factor from their data. If this correction is done **India's Youth Labour Force is much more than projected in the official/academic documents and therefore unemployment/joblessness is also correspondingly much more.** We need to address this matter because we cannot assume to go towards 2030 without all girls/women getting an opportunity to work for their own incomes (a basic need of women empowerment).

Definition of Work Force: All those persons in actual economic activity, be it on a daily wage basis or casual or regular or intermittent.

As a thumb rule, Labour Force minus Work Force is equal to Unemployed.

Definitions of unemployed/jobless: Jobless are those in Labour force who are not able to enter into any job or out of work for very long periods (years). Unemployed are those who have entered Work Force but are for a number of days/weeks in a year are without any gainful employment.

Work done men or women in their own houses without any pay is not considered as gainful work. If someone raises poultry and the product is consumed at home, it is not a contribution to GDP or gainful work. But if the eggs or chicken or duck is sold in market for receiving money, then it is gainful self-employed work.

No Gainful Work, No Future

While there are several components adding to the quality of life and happiness, the first and foremost is some wealth which for most persons hoping to enter the Labour Force and Work Force is some reasonable income, even if it were on a daily wage basis. After a start and passage of a few years, they aspire for some steadiness of income and a slow but steady growth of it.

During childhood, parents take care of the income earning roles. Large number of Indian parents even amidst their terrible poverty take care of feeding their children; tend to them when they are sick. It is a part of evolutionary process of spreading of one's genes/species.

During the long period (several centuries) of stagnation and subsistence level of the Indian economy, children learnt skills from parents for some economic activity and lived on. But with the growth of modern economy worldwide during the early part of 20th century, that subsistence agriculture based village economy can no longer sustain livelihoods. The colonial power introduced new methods. Post-independence, the Indian elites and rulers accelerated the modernisation process; a number of useful plans and projects were done to uplift the rural economy. The rulers, planners, elites and administrators were much more obsessed with their own upliftment. The workers in the organised sectors (a small percentage of the workforce) were inducted into such a socialist planning through several laws to protect their interests, but without concern for growth of productivity and increase of efficiency. Organised Work Force also became a part of the ruling class and was protecting its vested interests. The bulk of Indian people were left to drift and waited for the trickles to be doled out to them from the highly protected planned economy for the elites, with some crumbs thrown for the rural people. The political system adopted several slogans and occasional doles especially before elections to keep the rural people "opiated" en masse.

No doubt some of the world wide advances in health care and availability of cheap medicines as well as certain low level of food sufficiency obtained through "Green Revolution" avoiding famine deaths, ensured that people lived longer. Population grew in leaps and bounds to nearly five times more than what it was in 1947.

Taking the fashionable theme of Demographic Dividend being spoken in the developed economies which have obtained very long life expectancies and therefore nearly 30% of old in their population, our elite also talk about 'YOUTH POWER". They mix it with our earnings through IT export (albeit through low level value addition) and arrive at a conclusion that India will be a SUPER POWER by 2020 or definitely before 2030: YOUTH is the mantra, they claim. Then they go about prescribing more IIT's, NIT's, Innovation Universities etc., and also create a National Skill Development Corporation (NSDC) and enact Right to Education Act for making all Indian children to pass elementary school. Others find panacea in e-education, tele-education, Akash tablets etc. For those who are unemployed/ jobless etc., they create experiments like NREGA or Direct Cash Transfer through Aadhar (UID) Cards. Several thousands of crores of rupees are spent on computers, new administrative delivery systems etc.

But what about those youth who are now between 13 to 35 years? Do they get some sustainable income for themselves? Can they hope for a better future? Will their sisters/brothers who are under 13 years now hope to have anything better than them?

Let us look at realities through information available in some survey data given in the reports referred to before. **Let us understand the magnitude of the real problems faced by the** INDIAN YOUTH, then think of some pragmatic, sustainable solutions avoiding slogans.

Glaring Realities

"The unemployment rate (defined as percentage of the number of persons to the persons in the labour force) by five year interval groups shows that it is the highest among the younger cohorts especially 15 to 24 years age cohorts. In other words, the incidence or instances of those who are willing to work and available for the labour market but unable to find the work or employment is higher among the young (before 30 years) when compared to their seniors (30 plus age). It is the highest among the 20 to 25 years age cohorts. The situation of young job seekers in comparison to adults seems to be hard" (Quote from SMD/MV report). This is so with the women largely who are left out in the skew which is in all the statistics of Labour Force; if their percentage is considered, it is still worse. Unemployment rate is close to 8 to 9%. Also the wages are very poor for the youth.

A young man or woman in their transition from childhood to adulthood from 15 years to 30 years has to face the wretched life of 'no income', no job, no daily wage. What happens to their self-esteem? After getting bashed up in life until 35 plus age, they seem to get some employment; mostly in the unorganised sector.

SOME REASONS FOR THIS STATE of affairs for the youth can be seen from the extracts of the following information from the MSD/MV report.

Levels of education of youth workforce in India (2007-08 data)

- Only 4.9 percent of young workers had post-secondary level of education.
- Young workers who completed secondary level education were about 18.8 percent.
- 24.2 percent complete middle school education.

But school education is not all for an employable life, though it is a good stepping stone. Our policy makers and influencers seem to be satisfied chasing only statistics of literacy and school enrolment, college enrolment.

Let us further look at some extracts from MSD/RV report:

- the nature of labour market has been transforming from unskilled to highly skilled.
- Increasing demand for skilled labour and declining demand for unskilled labour, especially in the non-agriculture sector.
- Human capital means addition of literacy and education to the possession of specific skills by persons.
- **As per NSS 61st (2004-05) round, about 89 percent of youth have not taken any kind of vocational training and among the rest about half of them have received through hereditary practices. There is a negligible level of formal vocational training for the youth.**

One can easily decipher, especially from the above stated, why youth until the age of 30 plus are not able to enter the job market easily.

Of those youth who get some incomes from work (2004-05 data)

- 52.4 percent are self-employed
- 34.4 percent casual labourers
- Only 13.2 percent regular and salaried workers

Not much has changed during the past eight years. In fact economic slowdown has made things worse. Even in 2016 things have not improved.

How they can envision a future?

Further sector wise: (2007-08 data)

- 55 percent of young workers are in agriculture and allied activities (with much poorer incomes).
- 45 percent in non-agriculture.

Coming to the much hyped IT sector again from MSD/MV report) (2007-08) data

- The share of computer related services, a major component of broad IT sector, the total adult workers 15 plus age) was very much insignificant about 0.32 percent only. Within this, the share of young workers (15-24 age) is lower than the young adults (25-34 age) (MSD/DN inputs)
- Employment in the IT/ITES export industry is about 1.77 million in 2009-10 (CRISIL Report). It is better paying but very small in size.

Unemployment rate was the highest amongst the young graduates at 35.5 percent level. It may be more now. Poor non-graduate youth cannot afford to wait for any job, however demeaning and poorly paying.

The report also mentions about a new category of **Discouraged Youth** (those who don't pursue education and also left actively searching for job due to several past negative experiences). I suspect these numbers are rapidly growing. The greatest danger for the society though unscrupulous powers-that-be may use them.

- Average youth wages (including regular salaried and casual workers) was Rs. 57 in 2004-05. Some more information (MSD/DN report).
- The size of youth population in the 15 to 24 year age group is projected to be around 240 million in 2011 (3 fold increase since 1980's).
- Share of agriculture in youth employment declined rapidly than adults (see my comment earlier that the youth are fleeing to urban areas).
- India has 459 million workers in 2004-05. Of this 423 million (92%) is in the unorganised sector. Most of the youth are also in the unorganised sector. (The situation would have worsened further for 2011).

- Joblessness (27%) is much higher than unemployment rate. In absolute number, it is 203.6 million youth (15 to 24 years).

Some Forecasts (from CRIISIL Report)

- The population of 1.2 Billion in 2011 will go to about 1.5 Billion by 2030.
- For India to have a large work force (15-59 years) likely to swell from 749 million in 2010 to 962 million in 2030. (My observation, if we correct for women-skew, multiply these by 1.5).
- **If the current trends in India's labour participation and unemployment rate continue about 423 million in India's working age population will be unemployed or unable to participate in job market by 2030 (My note: Multiply by 1.5 if we assume women also would like to participate equally).**
- The workforce will increase all the more in states that are the poorer and offer the lowest employment opportunity.
- Engineering graduates churned out from Indian institutions will almost double over 2011 to 2015 from 0.37 million to 0.65 million engineers. Their employability will diminish further. (My note: If economic growth does not touch 10%, things will worsen).

Is there some hope?

I knew much of the above realities, through reading, discussions and actual field visits where poor people and youth live. But a research for writing this article, has sharpened my insight. (See also my book "Way Beyond The Three R's: India's Educational Challenge in the 21st Century, Penguin)

If the policy makers in the early years of independent India had listened to farsighted persons like Dr. N. Mahalingam who pleaded for vocational education at the primary/secondary stage itself, even below the current ITI levels, their job prospects would have been better and over the passage of a few generations the current youth would have been lot better. But our policy makers told him that they are emphasizing on diplomas to produce more engineers! We now see that even after 60 years most of our youth are dependent on skill learning from their families! !

Now the situation is much serious, because politicians and administrators are floating one or another populist scheme, which do not address the huge number at the bottom. NSDC for example is mostly urban high/medium tech oriented. Also some youth are attracted towards the "easy money" available through illegal or illicit work which will make them criminals. The discouraged ones (referred to earlier) may become anti-social.

Had the many items given in TIFAC's Vision 2020 reports and summarised in India Vision 2020 (Kalam and Y.S. Rajan 1998), been implemented holistically India could have transitioned from a huge low wage agricultural employment to large scale better income manufacturing and service sectors. Most youth would have been rightly skilled by such an integrated market pull. But it was not to be! India opted for shallow liberalisation and globalisation, keeping agriculture still in a low productive level (we could have had a 350 million production by now had the steps outlined had been followed).

Now coming to solutions:

Policy Items for Solutions

(i) Reform agriculture sector soon allowing for large scale contract farming including with FDI's.

(ii) Enable many PPP's to provide for micro-irrigation and drip irrigation infrastructure in areas which are only monsoon dependent.

(iii) Amend land ceiling and environment/forest laws to enable establishment of many (public, private and joint sector) industries near villages and village clusters. Even FDI's can be encouraged.

(iv) Remove restrictive labour laws such as laws restricting units that employ more than 100 workers from firing employing (CRISIL report). These laws impact employment in three ways:

  (a) Entrepreneurs are discouraged to enter industrial sector

  (b) Industries try to substitute labour by capital thus reducing employment intensity and

  (c) Creation of smaller-than-optimal firms to comply with laws which makes them non-competitive. Little surprise our GDP is not growing fast.

(v) Give special tax deductions for those businesses which employ flexible labour from youth population having low educational status.

Projects to be Done

1. Instead of just depending upon current NSDC alone, create at each taluk level, Skill Imparting Institutes for children right from age 9 or 10 years so that they are employable in some of the Industries that will get created as per policies above. Give them some English and Hindi training, so that they can get jobs anywhere. (I learnt this while discussing with unemployed youth at Coimbatore during 2005)
2. Create small Electricity Generating Units (1 to 5 MW) with all forms of raw materials – coal, wood or mini hydel etc.,) in most village clusters so that the youth can apply their skills and do some micro enterprise.

Above All

Let us not mesmerise ourselves as YOUTH POWER but really start attending to those who are youth now on a crash basis (13 to 35 years) and also do greater and better plans for those below that age. Let us deal with a few hundred millions of youth about whom I have given the statistics. Let us jettison WOMEN SKEW in the Labour Force projections and prepare all girls and young women for income earning work, even part-time!

Let us do everything possible to accelerate growth of GDP as well, so that more jobs can be created; more incomes generated; and therefore more taxes collected; thus spending more for the youth to be enabled and skilled for better income jobs.

Hope in the youth for a better future is essential to protect Nature and make a sustainable development.

❐

2

Plastics: Boon or Bomb?

Recently I came across a newspaper report that Supreme Court judge(s) observed that plastic wastes are the biggest bomb in India. I did not get into the details of the observations of the honourable judge(s). My intent is not to take any issue with such a statement which has been made in anguish seeing the report of the Central Pollution Control Board (CPCB).

But it is important for us to look at several other bigger bombs awaiting massive explosions. Many of them, in fact, can be considered as a great Boon, but being converted to BOMBS.

Top most in the list is the demography of our youth. We have written about it, in a number of articles. Also importantly in Chapter 1; there are many others who warn about it. But each year, we neglect several tens of millions of Indian youth to a point that they are misfit to any modern employment. We are happy to pronounce judgement that they are (even from those who get degrees) unemployable and add that Government has to take action to improve the "quality of education", "impart better skills" etc. There are others who create various surveys and indexes, flash them in media as to how bad our schools/colleges are but little concrete actions are taken on the ground. Most Indian youths are struggling to get jobs or decent wages.

Are they not potential big bombs? **A boon made into bombs?**

India has one of the finest and richest iron ore in the world. No wonder the first carbon steel in the world was commercially marketed from India's Deccan region, in the name Wootz steel which went to make Damascus swords. We sat on our ores after independence and sold some of them inefficiently through public sector export agencies. Now they have suddenly opened up mining sector. Except for a few actors who make the total value addition (from ore to final products like steel rods, flats etc.,) the mining is done rapaciously leading to serious environmental degradation, without adding any prosperity to the local or other Indians. Raw materials are exported instead of being utilised locally. Is it not the nature's BOON to India being converted to an environmental BOMB?

So is with COAL. So is with titanium, a wonder metal and many rare mineral resources, most of them being in rich deposits in India. The Chinese have converted their rare earth ores into a great commercial success and also a wonderful geopolitical and strategic tool, making big powers to come to them; bend to them!

Bountiful gifts of nature, with poor attention to investing modern technologies, come as horticultural products – vegetable and fruits. Still about a third of the annual production is wasted without human or animal consumption. Rural poverty mounts more and more.

Even with the relatively moderate food grain production which is hovering around 230 million tonnes per year, (which could have been easily 300 million tonnes or more), we have now excess of procured food grains about 60 million tonnes which do not reach people. On the contrary, it has blocked private trade, causing an additional inflation and disincentive to farmers. Are they not bombs which have been exploding periodically? A country blessed with a large part of arable land, reasonable water resources and excellent climatic conditions, struggles with food supplies, poverty (and sometimes suicides) of farmers. BOON made into BOMBS...........

One can add many more. In this chapter we will explore the role of plastics and plastic wastes. These can be compared to policies and actions that other countries practice for actual growth from their rich natural sources.

Plastics in Modern Economies

This section is referred to from "Plastics – the Facts 2012: An analysis of European Plastics production, demand and waste data for 2011" by Plastics Europe, an association of Plastics Manufacturers; report of the Committee to evolve Road Map on Management of Wastes in India (Ministry of Environment and Forests) March 2010. Some reports of the US National Environment Agency (updated as of November 2012).

Plastics are a wonder material which were invented and put to use in 1912 and more so after the 1930's. Germany's BASF had played a major role, with others catching up speedily. Towards the end of 20th century and the current one, it is difficult to think of any area of life without plastics. Safe and hygienic beverage containers, toys, furniture, automobiles, health care equipment of various types from injection tubes to other tubes for patient diagnosis, treatment etc., and whole sort of packaging industry. Plastics are ubiquitous in all sectors: Agriculture, Manufacturing and Services.

In the developed countries plastics make more than 12 percent of municipal solid waste stream (2010); it was less than one percent in the 1960's. US report says that inspite of such increase only 8 percent of plastics are recycled. But this is an overall percentage. For some types of plastics, the recycling rate is much higher. For example in 2010, 28% of PET bottles and jars were recycles (ref. US report).

Types of Plastics

The European report states that different types of plastics with a variety of grades are used to deliver specific properties of each application. Indian anti-plastics activists tend to lump all of them as one. We will address Indian wastes later.

As per the European report, the "big six" plastic types have large market share.

- Polyethylene-including low density (PE-LD), linear low density (PE-LLD), and high density (PE-HD)
- Polypropylene (PP)
- Polyvinyl chloride (PVC)
- Polystyrene solid (PS) expandable (PS-E)
- Polyethylene terephthalate (PET)
- Polyurethane (PUR)

These together form about 80% of overall plastics demand in Europe, which is about 47 million tonnes (M tonnes). Europe exports about 15 M tonnes of plastics in primary forms from Europe. World plastics production is about 235 M tonnes (2010) of which Europe is 21%, China 23%, Japan 7% and India lumped with rest of Asia as 16%. India's use of plastics is far less as can be seen in the data on India, given in this article.

Other salient data on Europe:

- Total plastic production of Europe was 58 M tonnes in 2010.
- Demand from converting/processing industry 47 M tonnes in 2010.
- Collected post-consumer waste about 25.1 M tonnes of this 10.3 M tonnes are disposed off and 14.8 M tonnes were recovered.

The collected waste is put to following uses: Recycling, Energy Recovery, Disposal/Landfill and Recovery.

They also seek people's involvement to realise that plastic waste is too precious to be wasted on landfill. Better performing

European countries ban landfill. The overall recycling performance by European countries vary between 15 to 30%.

The Data on USA

From the US report, the following can be seen.

- 31 M tonnes of plastics waste generated in 2010 (about 12.4% of Municipal Solid Waste – MSW)
- In 2010, US generated about 14 M tonnes of plastics as containers and packaging, 11 M tonnes as durable goods such as appliances, and about 7 M tonnes as non-durable goods such as plates and cups.
- Only 8% of plastics waste generated in 2010 recovered for recycling.
- In 2010, the category of plastics such as bags, sacks and wraps were recycled to about 12%
- Plastics found in automobiles do not come under this MSW report and are treated separately.

A brief good description is given in the US report about the recycling – from the user to recycled products.

As per the report about 1800 US businesses handle or reclaim post-consumer plastics. The primary market for the recycled PET bottles continues to be fibre for carpet and textiles; for recycled HDPE it is bottles. Efforts are also underway to reduce waste at the source of production of plastics.

The table in the US report gives the overall recycling percentage for various wastes and their absolute numbers (shown in bracket here as tonnes).

Waste		Recycling Percentage
Construction debris (1,291,800)	–	99%
Scrap tyres (18,400)	–	88%
Ferrous Metal (1,386,000)	–	96%

Wood/timber (343,800)	–	69%
Paper/Card board (1,252,000)	–	56%
Horticultural Waste (247,800)	–	44%
(Includes items used as fuel in biomass power plants)		
Food waste (703,200)	–	12%
Textile/Leather (159,300)	–	11%
Plastics (803,400)	–	10%
Total (7,269,500)	–	60%

Note: Not all categories are given in the above extract.

Coming to India

The MOEF document is not rich in data.

- MSW generation is assessed to be 0.1 M tonnes per day for the year 2001-02. About 36.5 M Tonnes a year. Central Pollution Control Board's (CPCB) study based on 35 metro cities, MSW generation is estimated for 2008 as 0.573 M tonnes per day that is about 209 M tonnes per annum. This is about 6 fold increase in 8 years.
- Average collection efficiency ranges between 22% to 60% Highest MSW generation is Kochi (0.67 kg/ Capita/day). Typical MSW has 51% organic waste, 17% recyclable, 11% hazardous and 21% inert.

There is an interesting statement in the report: "About 40% of all MSW is not collected at all and hence lies littered in city/town and finds its way to nearby drains and water bodies, causing choking of drains and pollutions of surface water". Note that these are not all plastics.

There are many recommendations and other observations on operational processes of garbage collections etc. Many well-meaning statements which are not followed, such as safety of workers and rag pickers should be given due attention with emphasis on education for using safety equipment and necessary immunisation. "State Governments must facilitate the making of Detailed Project Report (DPR) for even small towns......."

Plenty of material is written on segregation of wastes without realising the practicality for an Indian household. Indians still produce very small amount of garbage daily. They cannot store for a fortnight or more of their organic waste!

Other such statements are:"Stray cattle movement could be regulated and burning of garbage, leaves and other wastes be prohibited". Let us look at the capital of the country which claims that sufficient funding be provided MOEF etc., without even rough estimates.

Plastics Wastes in India

As per the CPCB estimate for 2008, plastic consumption in India was about 8 M tonnes per annum of which 5.7 M tonnes are converted to waste annually. (India is less than 4% of world assuming these above are produced in India). That is 15,722 tonnes of plastic wastes is generated per day. (Compare this with the MSW which is 0.573 M tonnes per day, most of which are not handled let alone recycled).

It is to be noted that the plastic waste estimation is based on an assumption that 70% of plastics consumed are wasted. Look at the European data given before; it is close to 53% only.

Further, the report says approximately 6,289 tonnes per day (of the 15,722 tpd estimated plastic waste) are neither collected nor recycled and find their ways to drains open lands, rivers, railway tracks, and wastes.

Thermoplastics (which are recyclable) which include PET, LDPE, PVC, HDPE, PP, PS etc., form 80% of the plastic wastes in India and 20% are Thermosets (not easily recyclable).

Commonly littered plastic wastes are polythelene carry bags, plastic wrappings, plastic plates, cups, spoons etc., are recyclable plastics. Gutka pouches, multilayer packaging, laminated packing etc., are non-recyclable plastic wastes. **The report states that in terms of types of plastics 90% of Indian plastic wastes are recyclable; only 10% are in non-recyclable category.** However, the report notes that unsafe recycling is being done causing health hazards, leading to production of products unfit for human and animal use.

The report also deals with international comparisons and **states that the per capita consumption of plastics, in developed countries is far more than in India.** The readers can derive the numbers from the data given using 300 million persons for Europe. The description is good. However, the recommendations get into "high horse" statements such as "undertaking mandatory responsibility of producers for R&D activities on plastic waste mitigation". The statement is without any idea of the degree of sophisticated knowledge base required for plastics R&D and the large investments. As noted in the report, our producers of recycled plastic waste products do so because they do not have the equipment and know how to do so. Most of them are MSME's, the neglected lot of Indian planners and governments.

Another statement "undertaking R&D for developing truly safe and biodegradable polymers…" these bombastic statements can be seen in the reports of the mighty S&T institutions of India and elite educational institutions for more than three decades! Nothing concrete has come about in the market, from India.

So let us not bury ourselves in such wasteful and time worn clichés. Let us look at the hard facts of Europe and US. They are also grappling with the problem to increase percentage of recycling. We can look at the Chinese as well as they consume a lot of plastics now. At the same time, whatever may be the moralistic statements we do from high levels or activists' shouts we do or take actions to deprive the poor Indians of the simple facilities of carry on plastic bags (we have written on it in the earlier book), plastic consumption in India has to rise much more. It will increase with the demands of growth and increasing incomes of Indians.

Benefits of Plastics to Agriculture

Europeans continue to waste food:90 M tonnes per year! Indians as they grow a little richer, waste too. Wasting of food is also an environmental hazard – much of it can be saved by use of plastics: especially the waste of vegetables and fruits.

In Europe 20% to 40% water is wasted. In India it is much more. Even in irrigated canals, the seepage is high. How can plastics help? See some quotes from European report.

- Plastic pipes can carry water over long distances for more than 50 years and also prevent bacterial contamination.
- Plastic drippers can help tailored irrigation.
- Distribution channels can be lined with plastics to reduce seepage losses.
- Water quality can be improved for industry and domestic uses by plastic filters.

There are several other applications to prevent soil erosion. Pipes for pumping water and connecting river basins etc., can be better realised through plastic pipes. Also cities can have better water supplies.

There are also applications for soil moisture conservation.

Above all, agricultural produce need plastics, when used judiciously to preserve the grains, etc., and prevent decay. The entire industry of plant nurseries depend on good plastic bags, saving the use of earthen pots which are heavy and also waste rich top soil in their making.

Plastics are friends of agricultural farmers and ordinary people when used well. These applications one day will pick-up when it becomes a trend with everybody hurrying helter-skelter, as it happens in India in most fields (the latest being the set up boxes for digitised TV distribution). The experiences of other countries in applications are before us. Let us prepare ourselves now itself. Let us use their foresight.

Plastic industries (primary producing ones and the waste recycling ones) are to be enabled to obtain necessary know how as technology transfer and testing laboratories. This is where the Government has to deploy it's already over funded S&T institutions and elite educational institutions. Let them be mandated (as the government gives them tax payers' money as funds) to address the concrete problems of plastics in agriculture, waste recycling etc., - not to do new R&D but to adapt the global knowledge to help Indian industries to grow quickly. We will need a least 1000 plastics recycling business in India, when plastics use picks up.

For addressing the problem of discarded plastics in cities, the solution is not in making more laws and regulations. Let us be practical. Let us organise through private sector (in PPP mode) or by NGO's etc., a simple plastic waste rag pickers' industry. They would get paid for just picking up any plastics waste on a daily basis. Don't make citizens or the rag pickers to segregate them first. That may be done in a separate plant in each city/town using labour and machines. This will also increase gainful employment in urban areas while serving a public good. Banning plastics mindlessly will only add to people's problems and economy will also be affected

Let us approach the issue of plastics scientifically and pragmatically. It will turn out to be BOON rather than BOMB, as plastic consumption in India has just started. The society and economy will need it, as much as they will need electricity and water.

❐

3

Maintenance: Mainstay of Sustainability

Sustainability is a buzz word which is used by almost everybody, the most powerful to the academic to the middle class Indian in his/her "intellectual" parlance. Activists and media use it in all forms – the narrowest one confined to CO2 emissions or solar power. The nuclear power temporarily occupied that position of sustainable clean power (as distinct from coal based or petroleum or natural gas based power); but nuclear power is in doldrums not only in India, but most of the world.

Many of the narrow "climate change" related "sustainability" prescriptions and mantras are also in trouble, as their economic sustainability was in doubt right from the beginning. But in a period of rapid economic growth, people were ready to pay taxes for the needed subsidies.

Now with economic down turn and struggling growth, "carbon tax" in Australia is going to be scrapped. Most European countries and even Germany have started slowing down on their "climate change" and "green" initiatives, which were burdens on their economy, thus being unsustainable in the long run.

USA never really took its rhetoric on climate change seriously. Now they are concentrating more on shale gas for electric power and energy needs.

Economic sustainability on a medium and longer

term, would be the main foundation on which other facets of sustainability have to be built.

Nature and Sustainability

But sustainability cannot be just left alone only to be tested for its economic or business sustainability. A bigger BANKER IS NATURE. It keeps on loaning to us its bounties, so that we can enjoy better life. But for any bank only withdrawal without any return to the bank, is dangerous; it will erode its vitals. Nature is strong and has many complex adaptive systems (CAS) to sustain even a number of plundering attacks on it. But there are limits. When the limits are crossed, CAS's fail and irreversible damages can occur.

If NATURE is irreversibility damaged, what we talk of as economy and society will be of little relevance. Irreversibly damaged soil will be a serious blow to agriculture and animal husbandry. Severely polluted waters can damage sub soils and flora and fauna. Too much overdrawing of ground water can cause droughts and diseases without water. Social controls may break down.

Excessive wastes from industrial and consumer products be it metallurgical, electronic, chemical, biochemical etc., may destroy many local habitats in cities and town.

We all have an individual and collective duty, to ensure that the limits are not crossed in our **drawing of resources from Nature.** We need to giveback to Nature, through various forms of recycling: For example, recycle the polluted water by purifying it and use it for our newer uses, or give it back to the ponds, rivers, forests etc., recycle the biological wastes and give back to soils as nutrition; (reuse the industrial products as newly processed/products so that they do not pile up on the earth. There are a number of commercially successful waste to wealth initiatives in India; but they need to be replicated on a massive scale all over India).

But above all, we need to experiment, evolve and adopt newer life styles which will REDUCE HUMAN CONSUMPTION. It is another important foundation for SUSTAINABILITY. **Draw less from nature per capita, not through the pain of poverty to many millions as it is happening now in India and many developing countries.** It has to be in such a way that all human beings enjoy good life and have all essentials of continual learning, easily accessible and affordable healthcare, all items necessary to have a good life for the progeny and above all an assured and safe old age.

Such a state will require a newer definition of good life for all: not just the replication of European, North American, Japanese etc., life styles and associated definitions of enjoyment of a good quality life. **Some of their elements will be good but we should weigh against criterion whether full adoption of it by all the 8 billion people on the planet uniformly, will leave NATURE still intact with sufficient reserves (as a good healthy BANK). It should not be a dead uniformity of consumer products but give options to diverse populations/individuals of the world.**

Most of the discussions on 'climate change', 'sustainability' or 'ecological balance', do not cover **this essential need to contain consumption,** on the lines described above. Most of those who talk or prescribe solutions, do not change their luxurious lifestyles of the developed world and ask for a near-total freeze of changes elsewhere: no touching of forests; let tribes live in forests but don't introduce newer technologies; abolish coal mines/coal power; no nuclear plants; etc. But such a position will not be accepted by others who do not possess much now, who do not consume much now (even for basic need of nutrition, health etc.).

The Reduction of Consumption Handprints on Nature is the Key Need for Sustainability

Having said that, it is difficult to address all that is required to achieve it. In fact I would appeal to all the elites, intellectuals, academics, activists, media and the middle class etc., to address this crucial issue, TRUTHFULLY. They will come up with wonderful solutions and be able to implement them because it is from them!

In the meanwhile, we will address in this article **one single item** relating to reduction of consumption: that is CONSERVATION of all what we have so that we need not draw from NATURE much more resources. **Let us throw out, the 'throw away culture'** which may serve the narrow interest of innovators and business houses.

Conservation

Many foreigners who visit India and look at its treasures of the past be it Taj Mahal, or temples or mosques or sculptures or paintings including many living traditions now, are wonder struck by the creativity of Indian people. Simultaneously, they are also appalled at the poor state of maintenance or conservation of these rich treasures. Our soils get degraded due to careless misuse; our rich biodiversity is not conserved.

Coming now to more 'mundane' items like lakes, roads, bridges, etc., in our day-to-day lives in villages or towns or cities, we find utter carelessness (callous as well). In the July 2013 issue of Kisan World Dr. N. Mahalingam under general title "Moving Finger" had addressed water issues for Tamil Nadu (TN). He had dealt at length about the conservation of existing resources be it the rivers within the state of TN or lakes and ponds. He had particularly, referred to the Tamil word 'Kudimaraamathu" – maintenance of lakes and ponds. Maraamathu in the key word: maintenance. Basically it is the maintenance of existing infrastructure to their full capacity and efficiency. When I was a young boy and living in a small village Kottaram in Tirunelveli district of TN, I have often heard this word "Maraamathu Laakkaa" i.e. maintenance department. In later years of my professional life, I found that 'maintenance departments' be it in government or private sectors are looked down upon as inferior. Normally those who are considered to be incompetent are sent to these departments.

Maintenance is the key to conservation. Why conservation is important for sustainability? Lots of natural resources are utilised (i.e. drawn from Nature) to create various infrastructures, water supplies, ponds, lakes, bridges, roads, waste management units, food grain storage places etc., If they become unusable, we have to create them again with additional expenditure of natural resources.

When it comes to conservation of biodiversity etc., they are products of millions of years of evolution; each of them have evolved under some special context and survived in those localities. Many herbs, aromatic plants, special seeds, indigenous varieties of cattle, even insects and bees belong to such categories. They may suddenly come to be recognised for some use, thanks to the modern researches. Countries like Germany preserve most of the genes of earlier agriculture, **in situ** forms. I am aware that a number of exotic varieties of rice in TN which were considered as delicacies when I was a boy, are no longer cultivated under the pressures of commercial cultivation. But it is easy to maintain them in **in situ** gene pools, and conserve them. Same comment will apply to indigenous cattle/birds etc.,

Regular Maintenance

Coming back to the 'mundane' items which are engineered products like railways, bridges, houses, machines etc., they are under different forms of pressures. As a people, we do not care about maintenance. We think we can 'penny pinch' and save the money to be kept for maintenance. If one looks at Drinking Water

Mission, new installations continue, every year with budgetary expenditure but much of the earlier installed ones do not give water and have fallen in disuse. Lakes are in bad condition and we talk of new projects like desalination of salt waters!

If we concentrate on 'Maintenance' of the goods and infrastructure that have been created right from the time of the British, we will save on several thousands of crores of rupees which may be spent by trying to create new services. Also we will save on enormous wastages that take place to our grain and horticultural products, if we revive and maintain the storage facilities already built in the past. Engineering wise, it is not difficult; one can maintain well and also add new improvements, absorbing many contemporary technologies. For example, if there were old air conditioners or fans in a facility, it is better to remove them as they consume more electric power; one can put modern ones which draw less power. Let us look at how Delhi, Chennai and Kolkata (many such) airports have been modernised using the old airports.

Dr. Mahalingam's idea of having the whole of TN networked (I will say 'tubed') with drip irrigation infrastructure is very vital; can be done as a mission mode through Industries. But subsequent maintenance is key to the success as the drip hole may be clogged by mud or salts in water etc. The controlling electronic/electrical system will need regular maintenance.

I am aware of many good water recycling plants failing to deliver after about five years or a decade, because little care is given to their daily maintenance. I have discussed this aspect of poor maintenance with many officials and engineers. One dominant theme which came up was: powers-that-be don't show interest in maintenance because the funds involved are small and are also distributed over a long period; whereas for a 'green field' new projects funds are very high and bulk of the commitment is done in the early stages of approval and therefore much easy to get their 'percentage' of money for various approvals.

This is an unfortunate reality in most sectors of India.

On the contrary, if regular maintenance (including disposal of old excessive energy/material consuming equipment with modern resource conserving equipment) are done, the capital-to-output-ratio of such projects will be low. It will speed up economic growth in the capital starved country and create a large amount of productive employment for millions of our youth.

Just imagine the drip/sprinkler irrigation network being set up in TN and other states of India and having a regular "maintenance force" on a daily basis. How many families will be uplifted in rural areas just through maintenance services alone.

Organising Maintenance

I had studied in detail many initiatives in India, past and recent past. After successful initial implementation, very few things reach their full potential. For example, the great initial success of Green Revolution, its spread to many other areas of the country as well as its sustenance in the initial areas, was poor.

It did not reach its full potential. So was the animal husbandry related initiative as we had described in an earlier chapter.

The entire primary Health Services and Second Health services including the tertiary one fall under the category – not being successful in sustained delivery of services to people despite major investments. India which started with electronics science technology and applications as early as late 1960's, when S. Korea had none, is now the most laggard in electronics industry including high value software generation.

India which was good in machine tool sector with HMT in commanding heights and a few private sector industries, is now dependant more than 80% imports. One can add many more. If we analyse the reason it will fall under following categories:

- The activity have been shifted under a regular activity of a government (State or central) department or agency. For example, the Green Revolution was made a part of ICAR's extension services.
- Or kept under the "Commanding Height" of a Public sector enterprise (PSE) as a monopoly: be it HMT, Coal India, FCI, ONGC etc., Monopoly breeds laziness (and also corruption).
- Opened to private sector but having too many controls and regulations without any understanding the ground realities of business processes (e.g. sugar industries, pre-1991 telecom or automobiles, electric power entities even after 2000-etc).
- Well thought out PPP models (plenty of examples); buzz word SEZ, FDI etc.).

We lack attention to details, we lack foresight even for a short term (let alone long term) and we lack **perseverance and patience to implement to maintain.** Our **panacea** is to declare things as Government's duty, now a days as Rights through paper Laws, be it a midday meal or maintenance of lakes or taking care of FCI godowns and open storages with some covers on the top or railways or irrigation canals. Courts are also happy in passing such orders and catching some headlines.

Realities of the world of economics and commerce (trade included) are different. It is competition and survival and self-interest which keep things going. Socialist, welfarist ideas do not succeed in delivery of goods/services to people, as we have experienced over six decades (China realised it long ago and instituted entities with market characteristics and therefore competition at local levels).

Maintenance systems in India should be thrown open to Private Businesses with competition. Big businesses may open chains of Maintenance Centres at local levels as is done for Retail outlets. **The laws and rules should be such as to allow them to run as profitable businesses.** They will (through market forces) learn how to share the profits through local out sourcing. They will provide brands which can assure quality of services. May be a few more Private Enterprises will emerge which independently evaluate quality of maintenance services and ACCREDIT them, giving confidence to the consumers (of maintenance services).

No doubt such multiple enterprises may increase the cost of maintenance from the unrealistic 2 to 3% of capital investment as a part of the allowed budget. Government (State and Central) should disband their own so called maintenance units and get services through such **Private Maintenance Enterprises,** which are to be recognised as Services Sector Industries.

Maintenance Businesses in India are as important as Retail Businesses. It is also worthwhile considering FDI in many such Maintenance Business Network, to bring in newer technologies, quality assurance and business practices.

Let us strengthen MAINTENANCE, one of the key ingredient of SUSTAINABILITY so that we draw less from NATURE.

❐

4

Identity Crisis: Learning from our Traditions

India and Indians seem to be going through a chaotic phase. There are several forms of crises taking place daily and most of which are due to our own making. Irrespective of official statistics and media analyses, prices of day-to-day commodities are going up. Several quick fix solutions for distribution of agricultural products and bans on some traditional products seem to be damaging the livelihood of farmers and also damaging a good base of export markets assiduously built up over the past many years.

Industrial sector is slow to grow and creation of new jobs continues to be a challenge.

On the other hand, media is full of trivia which create passions and divisions. One would wonder whether many local violent incidents are partly the result of such created passions. Many new identities are exaggerated and people block roads, burn trains etc. to assert their identity rights.

Identity Crisis

The country is suffering the consequences of various IDENTITY politics which have been developed since independence for electoral gains. There is no **identity** as an Indian; the linguistic states for a while, gave an **identity** associated with the language: Tamil, Telugu, etc. Even then Hindi speaking states were not one.

Even within the linguistic states, there were (are) several separate **identities.** Some of which were traditional ones dating back to several centuries of evolution of Indian civilisation and also many historical partitions by kings, dynasties and invasions by foreign rulers. There were (are) many divisions of castes, sects of religions (of Hindu, Christian, Muslims, Sikhs, Buddhists, Jains etc.,); the tribes who have maintained their separate **identities** through all such onslaughts of civilisational changes, have also many divisions between them; such divisions or separate identities were(are) carefully preserved by them.

The Constitution of India considered all of them, including a number of immigrants who have become citizens of India, as one, having a single identity as INDIAN CITIZENS.

But that was only in theory – in legal text books and discourses. Or in the calculations of the technocrats who design IDENTITY CARDS.

Reality of Indians is that their immediate identities, had been traditionally linked up with their villages, local communities, castes, religions etc., Not all of these traditions had respected the modern constitutional principles of equality, freedoms for individuals etc., Modernising forces, post-independence, have challenged many of these traditions; still many of them have their vestigial remnants.

Impact of Electoral Politics

But, the electoral politics as it developed in India, have played upon many of these past differences to create sharp **hate-identities** of "We" versus "They". These help in the constituencies for legislative assemblies and parliamentary constituencies. A slight excess of such **identity groups** in otherwise multicultural groups, can help an electoral victory (even with 20% votes polled) for a candidate "nurturing" such an identity. The earlier (historically) evolved harmonies (even with some tolerable tensions) are broken down for electoral conveniences; that explains the emergence of many splinter parties to attract IDENTITY POLITICS VOTES based on religious sects, caste, language, tribal affiliations etc. Superimposed on these **identities,** many other forces including corruption, play their roles. But these forms of identity politics have taken deep roots in India. There are several such identity crises even amongst big "national" parties which are supposed to represent a **pan India identity and have modern economic social and global outlook.** They try to dance to different tunes as required in some regions, or for all India level "identity politics". They also add "modern" jargons like secular, socialist, nationalist, reformist, equity, human rights, Hindutva etc. In actual practice, in terms of their actions and specific contents, there is a near total confusion as to what they mean.

Therefore many young Indians suffer from deep identity crisis as to what they stand for. The process of urbanisation and the rootless educational school system have confused the youth; they have not imbibed the traditional values derived from the centuries-long Indian civilisation and culture. At the same time, they are given a superficial exposure to the Western modern culture. Ideas of "rights" are given without the attendant need for responsible behaviour. Freedom is not understood as being closely tied to social responsibility and civic sense. "Work ethics" has not become a part of self-discipline. Instant gratification appeals to most of them as the sign of success in life. "Greed" replaces "enlightened self-interest" which is the basis of modern governance.

Such attitudes are derived from the political processes in rural and urban India, at the grass root levels, middle and leadership levels. For electoral success various forms of fear, greed, hatred and irresponsible hopes are generated in people, in order to keep them within an IDENTITY group to get block votes.

In addition, other cultural identities even from abroad are shown through electronic media, most of them being distortions or exaggerations of real life. In the name of fusion many hybrid cultural identity forms are explored. It may be difficult for many people especially for the youth, to discern the basic values. There is thus an all-round identity crisis in rural and urban areas. It may be called "churning" as well.

Governance Systems

It is natural that the governance systems are the creatures of or subservient to the political power, be they kings, emperors, dictators, or various forms of constitutional rulers (single party republics to multi party parliamentary democracies). Dharma, varyingly interpreted, formed the basis of rule by most Indian kings in addition to the principles and practices given in Artha Sastra by Chanakya. There were several other guides such as Thirukkural; lot of it is devoted to rulers, governance, wealth generation and distribution; it also weaves through all the verses the principles of Dharma (Aram in Tamil). There are many similar books, poems and traditions in all parts of India. Tribal societies are governed and regulated well through the tribal traditions, culture as well as fairy tales.

In later centuries diverse traditions of Islam and Judeo-Christian traditions had their impact as well.

Coming now to the present Indian government system, it is based on a written Constitution which derives a lot from British governance systems as they were practiced in India, and in Britain and also from American governance systems. They have derived some elements from the Indian traditions. The principles enunciated in the Indian constitution contain some of best modern ideas about humanity and can be shining examples for other countries as well. The constitution has been amended many times since its adoption in 1950. The changes reflect many of the political processes we have described earlier; therefore not all of them are in the long term interest of Indian Republic and Indian people.

For example over centralisation had not been the crucial feature of Indian governance systems, even under the reign of many powerful Emperors. Villages enjoyed considerable autonomy. To be sure various separate identities existed even within the small villages, some of them despicable as was the case of untouchability. As can be seen from the evolution of Indian culture, the Bhakti movements, the impact of Jainism, Buddhism and other sects of Hinduism, work of Christian missionaries, active emergence of Fakirs, absorption of ideas and practices of many specific tribal or native peoples' cultures, impact of modern European Renaissance and attendant world outlook etc., all of them had a place in India and amidst Indian people.

In that process, many syncretic traditions and symbiotic forms of co-existence emerged. Many of them are still continuing, in spite of continual attacks by identity politics we have described earlier. In fact these syncretic traditions, and tolerance resulting from symbiotic existence, form the essence of Indian civilisation. Every Indian language has an equivalent of the Tamil saying: "Yaadum Oorey; Yaavarum Kelir"(Each place is our place;All are our relatives). Acceptance of all religions and faith, is the strength of India. It is exemplified from the bhakti literature and works of saints like Kabir, Guru Nanak Dev, Tukaram, Jnana dev, Ramalinga Swamigal, to name a few. Even now, some of those traditions continue, though in some cases, high pressure advertisements by a number of contemporary religious leaders tend to mask the inner undercurrent of "acceptance of all" (not just tolerance).

At this point, it is worth looking at a quote from a book by an English ICS administrator Philip Mason "The Men Who Ruled India" (2002).

"It is not easy for honesty and pity to flourish at the court of absolutism anywhere. No one need be surprised that such growths are hard to find among Mogul courtiers and officials. But somehow, among the peasants, in their stony soil of fear, hunger and poverty, these virtues did survive. They were still to be found in the villages, if you went to look for them, three hundred years later, together with a warm hospitality and the dignity of people poor in material possessions but conscious of long ancestry. It is hard to understand how these things survived, but they did."

Mahatma Gandhi saw these very clearly and hence was asking for decentralisation to the village levels. After about 66 years of independence, these may be valid but they are being severely shaken.

Unfortunately the Indian governance systems, post-independence,have been moving more and more away from these traditions. Under the names of secularism, post-modernist ideas of individual liberties as articulated in Europe and USA, the overly centralised iconisation by the media, etc., tend to disrupt the beautiful harmony carefully preserved through the evolution of syncretic traditions and acceptance of symbiosis in India, as a way of life. **Idea-absolutism is being forced from Delhi and by media monopolies.**

There is a continuing over centralisation of many items at the Central Government level and at best some elements are delegated to the state government levels. The policies and procedures are increasingly framed with short term political motivations, and also using technocratic/bureaucratic approach. One-size-fit-all "pan-India" or "pan-State" models are pushed: like identity cards, direct transfer of benefits, single model syllabus, single entry exams, single type of lists for subsidies etc. The knee jerk controls on movement and sale agricultural products, based on such single-type models, cause lots of local damage.

The way of life of Indians, which has seen many such onslaughts through its long history, adapts to more acceptable local forms through mechanisms of corruption, adhoc local changes (which are strictly speaking illegal and which is beautifully described in Bengaluru "adjust maadi"!) Some years later, some of these are "exposed" as scams! Why in the first place adopt rules and regulations that do not fit in properly? Why not adopt governance systems which empower local adjustments in a consultative manner? Yes, in the process, the centralised power structures in the political parties and in governments will become less powerful! **Do we want powerful India and increasingly better lives of Indians at all levels of the social-economic hierarchy? This decentralised governance is not a utopia;** some European countries like Switzerland practice this governance system, a large extent USA practises it; even the counties in USA have power; individual communities enjoy their freedoms.

If this is done, it will greatly reduce the IDENTITY CRISIS we had described earlier. When empowered locally, the people will learn how to exercise power, and to live more harmoniously. The persons who fail to implement will be available locally to answer, not sitting in distant air conditioned rooms. The person present locally to implement cannot escape pointing fingers at "Melidam" "Ooparsey order lenapadegaa" (higher place; we have to get the order from higher authorities).

Indian Languages

Governance systems are important first foundation to avoid the identity arises spreading amongst our people because in the existing systems they feel that they are powerless; they have to catch "somebody-in-the-power-hierarchy" to get even simple things in their lives. They are thus ready material to be manipulated for some form of "identity politics" or "false-hope schemes". They become so gullible to fall for various schemes announced as their final "saviours".

But as we have noted earlier, governance systems are subservient to the political powers and the political processes. Therefore there is a critical need to change the political processes away from IDENTITY POLITICS and IDENTITY CRISES towards more tolerant syncretic traditions of India. Even sharp differences can be adjusted in symbiotic relationships, as biology shows us and as Indian civilisation has shown us.

How to effect such changes to the political processes? Not by another spate of constitutional amendments or enacting new Bills. No doubt some of existing laws would require changes and even jettisoning, as most of them have been enacted under the heat of identity politics or political number games.

These apart, real changes can sustain in our society, only when we re-learn to capture the essence of our traditions, described earlier rather briefly in this article.

If we look back to recent Indian history, the entire Independence struggle revolved around the recapturing of the Indian traditions and gave them prime importance in the modern context. Blending the strengths of our traditions, with modern knowledge of science, technology, engineering, medicine, economics and administration, was the key motive force. We see this in the writings of many of leaders: Jawaharlal Nehru, Rajaji, Ambedkar, Subramanya Bharati, Vivekananda, Maulana Azad and many others from all parts of India. Though in detail there may be many shades of variations, on the aspiration of such a blending there was almost a near consensus.

But post-independence, this aspiration was abandoned, to be repeated only during some special annual functions, but not for actual practice.

In addition to the hurly-burly of electoral politics, abandoning of Indian languages (which the British described as vernaculars, the language of the slaves) accelerated the speedy disconnect with the Indian traditions. Though in form, linguistic states that were formed and the languages used for local administrations were Indian languages, they were mere replicas of English – that too in complicated forms, not making much sense to the people.

At a time, when British Government has made it mandatory for all UK government offices to use plain English (http://central-government. government computing.com/news/gds-adopts-kiss-principle-or-is-it-rule-for-whitehall-english), we are complicating government's language in English and in turn the Indian languages are made to copy them as translations.

Languages are the repositories of and identities of culture. That is why India has a few hundred well developed dialects, though not necessarily having separate scripts. Many dialects are falling into disuse and may wither away. In addition, the major Indian languages from Assamese to Bengali to Telugu to Urdu (see the Indian rupee note) are suffering from major atrophies. Even when students are taught in the local language, it is in some "cloned forms" to adopt the texts of NCERT written in English. The social studies contain some "caricatures" of local histories. Local people are often dismissed as tribes or with some pronouns describing them.

The entire bulk of literature which describe dharma in story forms or poetry forms or other textual forms are dumped as items to be specialised during post-graduation. The entire folklore (Naadodi Paadalgal in Tamil) are relegated to some special art forms. They were the strengths of our people, reminding them in simple language, the realities of life, the responsibilities in life etc. Similarly epics like Ramayana, Mahabharata etc., and literary master pieces of Kalidasa or Silappadgaram, Manimegaliai etc., (and similar ones in Indian languages), Pancha Tantra stories, as well as works in Indian languages about other religions like Christianity, Islam etc., are kept away from our more "fortunate" children or even from "ordinary" children who study in Indian language medium, under some pretext or another.

We can save the Indian languages only when all our children who finish standard 10 to 12 are reasonably well familiar with such literature including folk tales at least in one Indian language. **This familiarity should be beyond mere passing of examinations or scoring high marks. It should be such as to kindle in them an excitement to explore more and share more.**

That brings us to another important point. The availability of such literature, folktales, prose etc., **in down loadable e-forms on a massive scale is necessary to sustain their interest.** Today efforts of a few individuals or groups, are giving cyber presence to Indian literature. **But it should be on a massive scale.** Government should take a lead. Instead of squandering tax payers' money in schemes for creating data bases (partly aimed as a "stimulus" package for our IT industry), such a cyber project for Indian languages will be lot more useful to re-invent Indian traditions through Indian languages.

Simultaneously efforts should be made to describe modern knowledge bases in Indian languages not as mere translations of complex English forms but as simple forms to reach people. Let us bear in mind the British Government's instructions on plain English. The official instructions in Indian languages should follow similar lines, with the strength of local traditional use.

Look at the Indian word "Lakshman Rekha", how much more rich, it is in contents and dimensions than "Red Line" used by President Obama for Syria. Of course English has its own such words: but rich in their context than which can appear in Tamil or Hindi translations.

Thus a massive movement is necessary, not just to talk in Indian languages or to introduce it as a compulsory subject, but to create conditions for our youth to learn the truths and traditional wisdom about life through our languages. Then they will gain a special identity for themselves; when taken collectively in a holistic manner that will be the Indian identity which has evolved over millennia. **Simultaneously, for growth of our languages in modern knowledge domains, good knowledge of English is a must. Capturing into the Indian languages, foreign words as necessary, will further strengthen the process of having all Indian people drawn into modern life with the strength of our traditions.**

Conclusion

If this is well done, even our politics will change. The force of our traditions imparted through our age old literature, will be an antidote against the poisons of identity-politics-hatreds; excessive greed; irresponsible tendencies for instant gratification; unnecessary fears against "others". That force of tradition will also imbibe many positive features in our people: such as evolution of syncretic traditions in modern forms; acceptance of certain symbiotic relations as a necessary part of life; enjoying diversity and above all the ability to adopt the great principles and pillars of ethics: reciprocity and empathy (we have described about them in the earlier companion book).

An India with a new face and all Indians with new strength, vitality, vigour and hope will find a new IDENTITY for themselves. They will also shine in the world with the strength of Indian languages and English.

❐

5
Nature and Modern Life

We started exploring various aspects of living in the modern world as it has unfolded in the twenty-first century in a series of monthly articles in Kisan World from 2010. We have covered a lot of ground on issues relating to the modern world, the Indian context and actions required for maximising benefits for All Indians. Our word all covered all Indians, present now and to be born in the future and also those who may make India their home. But it did not mean each one will get an equal measure of everything.

There are limitations from Nature. First of all, the immense diversity of Nature would mean attempt towards a single uniformity is not possible. One size-fit-all solutions will not work except for the beauty of the words appearing as slogans and hypes.

We have emphasized all through the series, the importance of agriculture (which includes other primary forms life like animals, birds etc.,) in human life. Whatever modern forms of lifestyles, agriculture will continue to occupy the primary position especially for larger countries like India. Fascinated by the IT sector and large money that is obtained from foreign (developed country) contracts as outsourced job orders, (employing only a small fraction of educated Indians) the country's policy makers and the middle class had built up a scenario of "Services sector" and "Knowledge Economy" being the panacea for achieving a "developed India". The benefits of liberalisation and globalisation in 1991, were virtually hijacked by the IT sector lobby, leading to grow neglect of Agriculture, Mining and Manufacturing Sector.

Developed India

The major national exercise done Technology Information, forecasting and assessment Council (TIFAC) which led to the book "India 2020: Vision for a New Millennium" by Dr. A P J Abdul Kalam and Y.S. Rajan, emphasised a balanced approach. It pointed out that agriculture and agro-processing have to play a very major role for India along with mining and manufacturing. It is sad that much of those specific action plans were drowned in slogans of "Services Economy" and "Sunset Industries" (which were the traditional huge employment generating manufacturing and mining sectors).

Under the slogan of "Services sector", financial sector grew jacking up GDP figures. Reality sector grew on this hype, with large assistance from the black money already available "freely" in India. Corruption grew by leaps and bounds and crony capitalism is a natural consequence.

Thus the concept of evenly developed India was the casualty. India has now a skewed development based partly on the hard work of *Kisans*, equally hard struggles of Micro-small-Medium-Enterprises MSME's who have to face corrupt Government machinery, terrible electric power shut downs and severe global competition from countries like China, Bangladesh, Vietnam etc., a few honest big companies, and rest of it through crony capitalism, crony imperialism (of nurtured imports ranging from Defence to gas to coal to machineries to electronics to software).

Governments, activists, media, "popular" intellectuals, some gullible middle class persons and politicians think that this terrible skew which leaves out a huge bulky base (BB) out of productive economic activity, can be corrected through a series of "doles" through enactment of Bills for universal education, universal employment guarantee or a compensation, universal food security through government supplied grains etc., - all through tax money.

If wealth is not created, it cannot be distributed. This simple law of life is conveniently forgotten. Net result is that Indian people suffer.

What Have We Covered So Far?

Agriculture: With the above in view, it was but natural that the first six articles covered various facets of Agriculture. The emphasis was on many reforms needed to the governance of agriculture; to free it from the shackles of government controls, some of them being knee-jerk reactions to market reactions; vote bank politics etc., More importantly we have also addressed the modern technologies that need to be adopted to reduce rain-dependence. The sixth article was about the role of Biotechnology and the need to adopt many of it and against the irrational activism that creates scare about GM crops.

Standards: The seventh article addresses a very important issue about the STANDARDS. This is forgotten or ignored by agriculturists, MSME's and even big businesses persons. Much of the bureaucracy considers such issue as another file or sending a delegation to international meetings. Politicians do not bother about them at all. Mostly some vocal activists and media, looking for spicy news, build slogans on this issue; some even catching the attention of honourable courts, who are not fully equipped to look at the totality of impact on all Indians. If Europe has adopted or USA has adopted it, then India should! It is a sign of a developed, modern advanced India. It is well known that a number of such activists are supported by commercial/ vested interests from developed countries. They lobby for everything: low emission cars; anti-GM; ban on pesticides; ban on some medicines; ban on plastic bags; new standards for home/local made wines; food purity etc., The regime of

World Trade Organisation (WTO) covers Agriculture; it refers to phytosanitary conditions. Under that proviso, many of India's agricultural produce can be denied access to other countries thus affecting Indian agricultural exports. They may even, some day, cleverly argue that Indian five-and-three-star hotels should only procure international quality produce as otherwise they are being subsidised by poor quality Indian local produce! Note, only very recently European Union countries raised some objections (sought clarifications!) about India's Food security programme and whether it will disturb world trade.

It is not our case that Indian products should not have tighter standards. But it invariably means higher cost of production and higher price. Can all Indians afford to pay, should be an important criterion. If the people of those developed countries evolved over many decades to such tighter standards, why should India leap frog, leaving a large number of Indians in the lurch?

As a classic example let those who can afford even to import their bottled water, do so! Let them not shout for tighter standards when resources need to be put in cleaning municipal water supplies to supply reasonably hygienic water for all.

There was also special article on Animals and Birds for health and wealth, appearing in this book as a chapter. This is crucial for agriculture and involves many standards.

I will cite two of my actual experiences in coordinating internationally on the emerging standards, amongst the many in which I was involved during my professional career. During the 1970's and 1980's many of the currently existing standards, rules, and laws regarding the peaceful uses of outer space (PUOS) were coordinated in the UN committee on it: UN COPUOS. It still operates. India through ISRO was an active participant; I therefore had an active role. We could see the clever ways in which the developed countries tried to push for new standards which will push back the gains made by countries like India and China in catching up with space technology.

For example they pushed for very strong tightening of standards orbital slot location in Geostationary Orbit (GSO).

This would have made the satellite control systems very costly and also such technologies were denied to India, China etc., who had, by then, nearly mastered the earlier technologies used by the West. But the idea was sold in a clever way to other developing countries saying that it is like parking in New York (where UN meetings took place and N.Y. was notorious for congestion). But in reality in those decades, GSO was nearly empty. We also equally countered their move by talking to the representatives of developing countries and exposing the hidden agenda, of keeping the technology applications only to themselves. There were many such instances.

Later during 1990's when I was with TIFAC we came across another interesting case. TIFAC was then successfully executing Sugar Technology Mission (STM). One day the Mission director J.J. Bhagatcame to me and said that he learnt from Sugar Industry sources that Indian Ministry of Health was going to accept a new standard for sulphur content of sugar, proposed by some world organisations and developed countries. It would put most Indian Sugar industry in jeopardy. TIFAC itself at that time was working with a few Indian sugar industries on technologies to make new sulphur free processes. But it would take time for all Indian sugar industry to adopt.

Moreover, the amount of sulphur involved was negligible. It was not of any harmful proportion, nor were there compelling reasons through scientific studies. Such geo commercial politics are not uncommon even now.

Also in India there was/is poor coordination between departments/ministries on such emerging issues. Health ministry officials just took it as Health issue and left it for the "Indian delegation" to deal with, without the backup studies on its impact on India. Fortunately STM/TIFAC which came under the Department of Science and Technology (STM) was alerted by sugar Industry; we had good coordination with Ministry of Food under which subject of sugar came. A strong delegation from India was sent and this proposal was postponed for future.

Among the arguments given was "why target sugar alone for sulphur? What about whiskey? One peg of whiskey has more sulphur than the whole day consumption of sugar in tea/coffee". There were other arguments as well.

On the other side, TIFAC's alert to leather industry during early 1990's about their use of azodyes which were carcinogenic and therefore would be banned by Germany and others, was not taken seriously by the Ministry. When the ban from Germany came into effect, later (It was published by Germany years in advance) there was a hue and cry! Indian media talked of Western conspiracy!

Yes some emerging standards are good. We need not oppose all of them. Standard makers (in its draft forms) always give a lead time for implementation. We need to prepare technologically and equip the existing industry and workforce to adapt.

The way we in India had ignored the dimensions of technologies and standards, during the GATT negotiations since 1990's till it came to final stages of WTO, will make history. It is in this context, foresight is crucial which was addressed separately in this series of articles.

We can give more examples. We have in several forumsalertedour intellectuals and public to be aware of this modern tool called STANDARDS, which are used for domination of markets. Let us be alert, and use realistic standards for Indian people in which they can have confidence. Let us be closer to Nature, in doing so. This alertness and having our own level of standards are crucial for modernising India and all Indians.

Basic Issues about Living with Nature and Dilemmas

Chapter 7 in the companion book "Basic Issues and Dilemmas", is still valid and I would request the readers to revisit it. Living with Nature in the modern forms is a great challenge. We will all be bombarded with new knowledge, fast changes, severe competitions, and unforeseen disasters (including items like financial sector, Wall Street, collapse of 2008, the consequences of which are felt even now). In addition hundreds of TV channels and trillions of bytes in social cyber media hit us. Terrorism is becoming endemic. Similarly many viruses from HIV, H1N1 to avian flu to several more variants are causing serious deaths.

Population is on the rise, due to the totality of modern facilities, as longevity goes up and may soon reach 100 years for those who are born in the 21st century (born after 2000).

Finding some reasonable answers and to have a balanced life for all, is the challenge, we need to face and find practical solutions. **In fact, the entire series is meant to equip the readers with a mental make-up (mind-set) to find solutions all through their lives.**

Disasters and Nature's Perils

Chapters 8 and 27 of the earlier companion book were on Disaster Management. As human societies grow in size and get more prosperous, the impacts of (losses due to) natural disasters and human induced disasters become huge. When we are struggling to build up infrastructure, large portions of Uttarakhandis being washed away. Orissa and parts of coastal Andhra face regular and severe cyclones. When the human kind assiduously tried to control CO_2 emissions from its industries, Nature burns many thousands of hectares of forests and bushes releasing many tonnes of CO_2 to the atmosphere, and also causing damage to human and animal lives.

Dangers to humanity and infrastructures built for modern life are not just from earth alone, there are cosmic sources as well. These dangers are described in chapter 27 of the earlier book. Such perils from Nature can perhaps not be eliminated by human beings even after several centuries. But humans can learn to anticipate and build up mechanisms to minimise damage. Building such capabilities against "external dangers" can hopefully, also provide a platform for the human kind to work together instead of being at war.

Globalisation and Related Issues

The world has a reasonable experience in the fast moving Globalisation of today for about three decades. It was heralded as a WIN-WIN for all by the free-trade protagonists. While it has brought in a number of benefits for many countries including India, (the fast growth of recent decades was due to that) there are serious other issues which are worrisome. Stating this is not a case for getting into isolationism or controlled economy of the past (44 years after independence). Globalisation of economies/ markets brings in concerns of other parts of the world: how they look at modern life, how they look at the relations to Nature etc. Also the competition sharpens. The efficiency demanded of Indian economy and therefore of productivity of Indians come into sharp focus. **In a globalized economy and market place, those who are less efficient and of lower productivity lose out.** Part of the current problems in India (in sharp contrast to China) is due to the fact that India ignored many issues which are addressed in this chapter.

An immediate associated issue is about the knowledge asymmetry between the developed world/developed companies and developing countries/laggard companies. This knowledge asymmetry is pervasive in India in polity, business, academia, "intellectual space" etc. We have to learn to look at the mirror and rectify the problem instead of being complacent that we are a great "knowledge super power".

Learning

Continual learning will therefore be crucial for **all** segments of society and all sectors of economy. Without such learning there can be no innovation; no ability to take advantage of modern knowledge; no capacity to live with Nature, drawing less from it and gaining more benefits for all Indians.

Chapters 11,12, and 13 of the companion earlier book address various aspects of such learning societies with strong emphasis on the Indian context. In its chapters 16 and 17, specific emphasis is given to the preparation for ushering into a continual learning paradigm in the Indian economic sectors: agriculture, manufacturing and services. Specifics given in this chapter address immediate, short term and medium term actions with the long term view of living with Nature in the modern forms.

Governance

All what we can understand, learn, know and aspire about India can be achieved only with a proper form of governance. There are several facets to it. These are addressed in chapter 14 "Governance Creative and Adaptive"; chapter 21 "Governance; Essential and Critical"; chapter 24 "Corruption: Corrosive and Cancerous" of the earlier book. In chapter 1 of this book "Youthful Indians: What is future?" we have also discussed solutions for having good governance.

The emphasis based on the empirical experience of India over the past six-and-half decades as well as other countries, **is one of strong decentralisation and Learning from Biology and Evolution.**

There is a chapter 19 on Ideology of Diversity: Role of Youth and one on the Values for a Sustainable World in chapter 25 in earlier book.

The earlier chapter of the present book is about the Identity crises engulfing India and how to restore a balance through our traditions and languages. This is a larger issue of governance and polity.

Miscellaneous

Another important element for modern life is access to

affordable medical services. This is a crucial issue. Copying the West is of no use. There have been successful experiences of mobile diagnostic centres in India. They will increase access many fold with very low investment costs. There is no question of having modern life for **all** Indians, without reasonably easy access to modern medical services. In a similar way, unique solutions to India need to be found for Waste Minimisation and Utilisation, Use of Plastics as well as acceptance of and facilitation of organic growth of slums with reasonable access to modern life. In a similar way, India would have to shun the "throw away" culture promoted by the Western societies and seek its own methods for Maintenance as a mainstay of sustainability. Maintenance as a profession needs to be respected. A large workforce needs to be skilled for it.

For the modern life multiple changes within a generation is a reality. Therefore the capability to foresee through modern tools of foresight (TIFAC is one example) is crucial; but foresight does not mean just to give a "vision" and forget the details. Foresight and attention-to-details need to be integrated.

We have addressed these items in the earlier and the present books.

The world is churning. Geopolitics due to the compulsions of domestic economy, job losses, availability of a new energy sources like shale gas as well as decades long frustrations, fears and hatred which were building up due to some systemic problems, serious corruption and opaque banking systems are very serious too

It is not easy to speculate on them. Indian media – TV channels, newspapers, cyber media and even social networks - do not care about such issues. We are too engrossed in some local micro – micro issues alone: a murder, a rape, a scam, a CAG Report, an accusation by an activist etc. Not that they do not have a space for consideration. But we do not even worry now about the border with Pakistan and the impending elections in Bangladesh, Nepal, Maldives or Myanmar. Most Indian elites won't even think of Myanmar; rapid changes are going to be there, challenging much of our low end trade. Do we convert challenges into opportunities or a potential disaster which strikes like a cyclone because we did not care for the early warning?

On the other end of the spectrum, military strategies in the developed world have transformed radically. To quote General William L Shelton, Commander, Air Force Space Command USA. "Our military satellites are technological marvels providing time-critical global access, global persistence, and awareness. These systems not only provide foundational game-changing capabilities for our joint forces, they also have become vital assets for the global community and our world economy".

Thus Indian and Indians need to live in a such a world of geo-politics churning local politics in other countries impacting others including India, global commercial competition (fair, unfair and clever), high levels of military capabilities already achieved or being achieved/access by other countries as well as various forms of terrorisms being witnessed within India, with terrorists achieving almost a walk-in type of capability to strike the places of their choice. That they do in Europe as well cannot be used for our own inaction.

Indian elite, intellectuals, activists and general public, cannot afford to ignore these dimensions of life. Living with Nature does not mean a passive, "green" and slow moving life. It also demands the ability to be aware of and agile for all that is happening and be prepared for what may emerge. It requires a special vitality to enjoy Nature.

Nature, Collective and Individual

Having discussed all the above, let us remind ourselves that they have to be in the human context. Humanity has to exist as a part of Nature! But a theoretical Nature, as an invisible invocation when human beings have to sacrifice their good lives cannot make sense. All the earlier socialistic rhetoric of collective good as against individual good, has boomeranged. Now people, especially youth, are strongly oriented for individual "good", quite understandably. **If individual as a unit is a forgotten collective, it is vacuous. If collective is completely ignored for instant individual gratification, collective will collapse and individuals will collapse as well.**

This is where the "laws of Nature" comes in. There has to be a balance between human individual and human collective; and well beyond the super collective biospheres earth, oceans, atmosphere, space, outer space etc., which is what Nature is, Nature is not a mere laundry list of such items. It is the sum total of these and many more which we do not know now, but will know later. Our idea of Nature itself is an evolutionary, growing one. Similarly the collective and individual are in symbiosis.

It is difficult to find firm and final answers. But posing questions and searching for answers, will help individuals, and the collective of individuals.

These issues are discussed in the earlier book chapter18 "Nature Common Good; Global and Local" and in this book in chapter "Individual and Collective". The relationship between Local(s) and Global is somewhat similar to relationship between Individual(s) and Collective.

I would request the readers to ponder over these questions.

Further On

We had attempted in this brief chapter to take stock of what we have covered so far in the earlier book and the current book so far. Naturally many of the chapters addressed immediate issues for better life for **all** Indians. They therefore addressed the nitty-gritty of government. policies, technological options, skilling of Indians at various socio-economic levels (Bulky-Base, Lower Middle Class, Middle Class, Upper Middle Class, Rich and super Rich) etc.

Often times in these chapters, we had to appeal for a mind-set change from the current one-size fit-all centralised solutions and governance systems **towards decentralised, diversity-accepting-governance systems.** This is crucial to succeed and to give good life for **all** Indians (in fact for **all** people of the world if their countries try to apply these principles in their contexts).

This conclusion and appeal are not only based on the human empirical experience in many countries so far including India, but is also based on modern understanding of biology and Nature: **Diversity is the essential guiding principle. Life is not a mass production line which turns out 'uniform' products with six-sigma-quality.**

Continual learning from the coded – messages of the past and the current experiences as well as continual adaptation is the essence of life. There is competition yet a balance; attempts for predation but self-limitations to that.

Of course, we cannot, within a human society, adopt some of the ruthless punishments which Nature gives. But we need to create mechanisms for self-limitation and self-control for individual and collective levels.

We can learn a lot from Nature about symbiosis and syncretism. Many older civilisations like India have successfully lived/live with them for centuries. Some of these are being threatened now either through some sub-collective Identities or individual extremisms.

We hope to address some of these in the forth coming chapters of this book.

❒

6
Precision Agriculture

The overall theme of this series, living with Nature in the modern world would also require addressing of a few futuristic items with high science and high technology contents. A few of them relate to the use of biotechnology; but many of these applications are prevalent on a fairly large scale in the world and on smaller scale in India. We have addressed them earlier. But precision agriculture (PA) is really a futuristic one, in terms of large scale adoption, even in the developed countries.

It is useful to be aware of the developments in PA in advanced countries. There are some persons working in this field and are publishing papers. I had not personally visited any agricultural field in India or abroad using these technologies. But my experiences with remote sensing, meteorology and GIS, GPS, automation etc., were of great help to understand PA. There are a large number of articles on PA available in the website, many journal articles as well.

Definitions and Description of PA

One such article from Wikipedia, creative commons, gives a few descriptions/definitions of Precision agriculture (PA). Other names used for PA are satellite farming or site specific farming (SSCM). "It is a farming management concept based on observing, measuring and responding to inter and intra field variability in crops. Crop variability typically has both spatial and temporal component which makes statistics/computational treatments quite involved."The article goes on further to describe the details as well as the problems:"The holy grail of precision agriculture research will be the ability to define a Decision Support system (DSS) for whole farm management with the goal of optimising returns on inputs while preserving resources".

However, some major companies like IBM consider such computational challenges as an opportunity for future business. From a brief article by IBM in the internet one notes "For IBM researcher and distinguished engineer Ulisses Mello and a team of scientists from IBM Research – Brazil", the daunting challenge of feeding 9.2 billion people for the world in 2050 "lies in real time data gathering and analysis. They are researching how "precision agriculture" techniques and technologies can maximise food production, minimize environmental impact and reduce cost.

Their definition of PA is "by collecting real time data on weather, soil and air quality, crop maturity and even equipment and labour costs and availability, predictive analytics can be used to make smarter decisions".

"Ulisses envisions a day when smaller farms and cooperatives could use mobile devices and crowd sourcing to optimise their agriculture."

In a book by the Australian Centre for Precision Agriculture called "A General Introduction to Precision Agriculture" (www.usyd.edu.au/su/agric/acpa) an excellent brief history of PA is given along with definition.

According to that book, impetus for the current concept of PA in cropping systems emerged during late 1980's. It was a "technology push". The newly developed Variable Rate Application (RA) equipment for fertilisers could be used to apply fertilisers at ratcs designed to complement changes in soil facility (SF) maps. Such SF maps were created with grid-based sampling of soil chemical properties. At that time Crop Yielding Monitoring (CYM) technologies were only at research stages.

During 1990's CYM devices started hitting the market. The satellite based NAVSTAR Global Positioning system (GPS) started to be available around 1990, albeit with limited capabilities becoming fully operational by 1993.

The book chapter further points out that the grid sampling at the intensity required to correctly characterise variability in soil and crop parameters proved cost prohibitive so by late 1990's 'zonal' management approach was used.

Further on, as per this report, the advances in Global Navigation Satellite system (GNSS) technology (1999) enabled guidance of farm machinery, auto-steering and controlled traffic farming. Advantages of CTF are minimisation of soil compaction, minimising input overlap, improving timelines of operations and reducing driver fatigue.

Thus, since the first decade of the 21st century there is a swift rate of the adoption of PA. The book points out about many definitions of PA and their choice of two.

One from the US House of Representatives (1997) PA is "an integrated information and production based farming that is designed to increase long term, site specific and whole farm production efficiency, productivity and profitability while minimizing unintended impacts on wild life and the environment."

As per this definition, PA is a "whole-farm" management strategy and not just for individual fields. It therefore applies to other agriculture production systems like fisheries, animal industries, forestry etc. An example cited in the book is: the tailoring of feed requirements to individual milkers depending on the stage of their lactation in dairy enterprises.

The second definition in the cited book narrows down PA to **timely management of variation** down to its implementation in cropping systems: Site-Specific Crop Management (SSCM) is a "form of PA whereby decisions on resource applications and agronomic practices are improved to better match soil and crop requirements as they vary in the field".

The second definition is much more focussed on the site and thus relevant for implementation by individual farmers. While there is a major application of sensors, communications and computers, modelling etc., **it needs to be strongly emphasized, that PA or SSCM is not just use of Information Technology (IT) to agriculture.** In India we have a tendency to reduce every modern knowledge intensive action to IT application, as we reduce them to jargons like Identity cards, Smart Village etc. **The focus of PA or SSCM is resource-use.** This requires strong rooting in the domain knowledge-bases relating to agriculture. For example:

- Optimising production efficiency of crops
- Optimise quality of crops
- Minimise environmental impact
- Minimise various risks due to weather, pests etc.,

There are also some attempts to link:

- Input supply chain management (Fertilisers, seeds, water, pesticides, labour, machinery etc.) and
- Output supply chain management(Harvest and post-harvest elements up to the retail or export markets).

Adding these may be ambitious to begin with as SSCM type PA itself is quite complex.

Implementation of SSCM/PA

If there were no spatial or temporal variability in the farms (fields) then the use of PA techniques is reduced to one-time extensive measurement of soil parameters and crop yields for various crops, and **have a look up table** for uniform application of fertiliser, water, etc., and timing during the different stages of plant- phenology.

But when variability is seen spatially and/or temporally, monitoring of the variabilities and doing the necessary modelling to optimise various parameters for better production, lowering inputs, reducing environmental impacts etc., are necessary; but it cannot be a mere technocratic exercise as the actions have to result in better profits, better "return – on – investments (ROI's). Therefore as the book points out **"the observed magnitude in the variability should be related to a bench mark level below which it would be uneconomical to attempt to manage".**

As of now such economic calculations will be based on a short-term economic perspective. It is difficult to quantify the environmental benefits to a quantity in a fiscal sense. May be perhaps in a long term future, it may be possible.

SSCM should thus be applied in an evolutionary manner, taking into account the cost factors and return on investment. It is explained through a figure in the introduction of the book.

1. First step in Uniform Crop Management (UMC): This will involve bulk or composite soil and crop sampling. As a part of the Vision 2020 follow up by TIFAC (Technology Information and Forecasting Council), the late Prof S. K. Sinha and his team of agricultural scientists applied this system in Bihar, UP, Uttaranchal etc., The land holdings were of small sizes ranging from 1 acre to 1 hectare. A few samples of soil were taken from each of them to characterise them. The spatial density of sampling may not have been high to qualify for classifying as PA, but it was probably the first time when the individual farmers fields were measured for soil parameters. Then using the agronomic knowledge of the scientists and through several consultations with the individual farmers collectively and individually, advice was given regarding the type of high yielding seeds to be used; the

mix of fertilisers, and other micro nutrients to be added, etc. Care was also taken to ensure that the crops suited the demands of market. Prof. Sinha had also advised farmers not to follow each other likes heep and instead use different varieties of crops so that there is no glut in the market due to increased production.

The production increase went up by 2 or 3 times (as those lands were cropped earlier on subsistence agriculture mode with poor quality seeds). Though the newer high yielding seeds per kg costed more, the overall input costs of seeds, fertiliser, mineral etc., were less than before as they needed to buy lesser quantities of seed, optimised fertiliser mix etc. The ROI for the farmers was very high. It was a great success story which expanded further.

The late C Subramaniam (CS) built up the National Agro Foundation (NAF) with TIFAC assistance, with a similar idea of detailed soil characterisation. It is doing well now, serving many farmers. NAF's operations are growing. CS and I had a vision of having at 100 NAF's all over India. Even these few examples are nowhere close to the early actions required for this first step of SSCM described above. But they are good beginnings.

2. The second step is an Interim Step where stratified random sampling within a zone is done. The number of samplings will depend upon the expected spatial variability in soil and variability for various crops. These are considered as the initial stages of SSCM (or PA).
3. These experiences have to lead to the **Future Approach** which is the **Third Step,** where site-specific-crop and soil management will be done involving fine grid sampling or sensing/scanning in the spatial domain. There will also be temporal elements introduced including crop related data and variability in weather etc.,

In considering practical aspects of SSCM, it is difficult to cover in a short chapter all parameters of measurements, related sensors, and other details of sensing and delivery equipment which may involve selective robots as well.

We will illustrate a couple of examples: The GNSS system which is one of the core enabling technology for SSCM (or PA) is now available in many forms. Receivers with accuracy in the range 10 to 20 metre would cost around $ 200. Those with accuracy in the range 2 to 3 centimetre would cost around $ 60,000. ISRO has recently launched India's own Navigation satellites named IRNSS. It will take a few years to give full range of services, say by 2017.

A whole range of sensors and monitors exist in the world market: yield sensors, biomass and crop response sensors etc. (see details in the referred book chapter). Then comes a range of functions like **Attribute Mapping** to filter out of the multitudes of raw data to derive the right user friendly information. Then come **Decision Support Systems (DSS)** DSS may have to be tailor made for different places. Then comes the application of VRA technology for delivering fertilisers, water etc.

Relevance for India

There are a few newspaper reports on PA in India. One by Anil Rajvamshi dated 15th October 2013 in the New Indian Express, gives some details and also raises some pointed questions in the Indian context. Another brief report dated November 19, 2013 in the New Indian express mentions about a Workshop held at Hyderabad 'Strategies and Tools for Adoption and Implementation of Precision Agriculture'. There is a paper in Map India Conference 2004 by U.K. Shanwad et al. It is overly optimistic. There is also a good paper on Precision Farming by Dr. Anil Kumar Singh, Water Technology Centre of IARI New Delhi. They do not however (their focus is different) make a full review of various elements and their status in India. It can run to a report of several tens of pages, if done systematically. A compilation of presented papers in a Workshop will not serve such a purpose of overview. In my understanding Indian agriculture is still far away from PA or SSCM though a few small scale experiments may take place in some larger farms and in Agricultural Institutes. It is useful to do these researches to gain experience.

A very large percentage of Indian agricultural fields still do not have systematic measured data about the soils and crops grown in them.

Instead of waiting for a much hyped Mission by the central government, it is important to start such measurements and data collection at several local places. One does not have to wait for super-accurate geo references. Google earth is a beautiful example of giving useful data as a guide map to local places. Students/teachers of 35,000 odd colleges spread all over India, can be mobilised to collect soil/crop data locally with field references mentioned (as it is done in the village).

Mobile phones used locally in the field scan send the data to the college. These data can be utilised to model and to advise on crop management. Aim should be to cover all agricultural lands in India. Agronomic advice should be a network of scientists from State/Central agricultural institutions. The CSR funds, which was explained in the last chapter, can be usefully deployed for this purpose.

This is the first base line data for moving towards better scientific management of Indian agriculture, to reach individual farmer.

Then probably over the next 40 to 50 years the subsequent steps (First, Second, Third steps described earlier) can be mastered. In the evolutionary period, some elements of high resolution remote sensing can be creatively used. There is no need to attempt an All-India or even an All-state grid. **SSCM or PA means moving towards individual plots.** The cost-economics should decide as to levels of zones or sizes of plots of sampling rates. But these should not be done as mere technology pushes or demo or a pilot. We have too many such 'science-pushes' which hardly help the farmer in the field to get better profits.

In spite of three decades of hype on Agromet services to the farmers (starting with huge hue-and-cry over the Cray Super Computer import being denied to India), the benefit to the Indian farmer even with village level medium range rain weather prediction, is eluding. Without good medium and short-ranges weather services up to the farmer level, dreaming of PA will be an incomplete exercise.

We hope that some capable readers of this book along with some advanced agricultural institutions and other engineers/ modellers, will take up the challenge of **drawing up of a detailed plan and road** map towards scientifically modelled and profit making agriculture for Indian farmers. It should not be a science or technology push but a practical guide to the farmers of using relevant technologies and models profitably. In that process, natural resources also can be conserved and environment can be better protected. But the bottom line is the profit to the individual farmers. I am not unaware of the policy, procedural and governance hurdles that impede the profitability of agriculture in India. We have addressed these issues in a number of earlier chapters.

Even when we were planning to write this article in the Kisan World on PA which is on a long term, futuristic matter, I saw a report in Business Line 16-12-13 titled "Many Farm Commodities selling below last year's levels". Of course, the economists have many explanations such as "moderating trend" or a "bear cycle". Even if these were true, one should be able to model them and work out suitable insurance systems and future trading systems to cover the market risks of the farmers. The aim should be to make agriculture (plant related, animal/bird related and fishery related) a profitable business giving better and better incomes to the practitioners. It requires a modern mind-set and corresponding governance systems. In the interim phase of transition, the farmers should have a security cover and adequate training for transition.

Ancient Wisdom

When I was studying various papers before writing this article, I was reminded of the chapter on Agriculture (Uzhavu) in Thirukkural. Two kurals capture the essence of PA. I could not resist my temptation to quote them.

1. Kural 1038:

   "Eyrinumnandraaleruviduthalkattapin
   Neerilumnandrathankaappu"

   A free translation is "ploughing, manuring, weeding, right water-supply and regular watching each one is better than the other". Implied in it are the careful optimising of such parameters. Today due scientific and technological advances we list more items and use PA.

2. Kural 1039

   "Chellaankizhavaniruppinnilampulanthu
   Illaalinoodividum"

   Free translation: The owner of the field should go daily to the field to watch and tend. Otherwise as a wife sulks with her husband (who seldom comes home), the yield will be reduced." What a succinct metaphor. SSCM and PA are meant for continuous watch on the field.

Both free translations are the English rendering by me from Kavignar Sirpi's Tamil rendering of Thirukkural. Let us keep the essence of the ancient wisdom in mind when we try to evolve PA for Indian farms.

❐

7

Soil: Key for Sustaining Life

When I started writing Precision Agriculture (PA) I was in parallel working on Soil Biology as well. One of the basic elements of PA is constant monitoring of soil. Beyond the usual understanding of the top conditions of the soil, its inner structure, the ecosystem it contains are not even taught to the children. Soil is taken for granted until it creates a problem. Since the PA article was long, I decided not to refer to these and came to a conclusion that SOIL in itself, its biology and their conservation deserve a full article.

I was delighted to receive a letter from Kisan World, enclosing a letter from a reader of my articles: Shri GANGASWAMY, retired Deputy Director, Horticulture, Bangalore. It was a New Year's (2014)gift to me! His suggestion was to address Soil. What more can I ask for! I am grateful to him. Some extracts from his letter:"Soil is the basis for all living organisms". In an article in the National Geographic Magazine the author takes a cross section of the earth which looks like a slice of cake where he shows millions of living organisms which harbour in the soil. I had been made alive to these under-soil organisms and ecology by an article in the Scientific American about three decades ago.

Shri Gangaswamy also points out how the first thing which the government of independent America did, about 200 years ago, was to take actions to conserve soil. He emphasises that the foremost duty of the independent India should be to conserve every inch of soil.

We fully agree with you Gangaswamy. There can be 'no living with nature' if we do not take care of and enrich whatever soil we have now. Even the planetary exploration, be it to Moon or Mars, lays emphasis to learn about their soils: any future human habitat will depend upon how good the soils are, to support life.

Land Resources

In our living culture, we Indians, tend to ignore the preciousness of water; and to soil we give a low position. The middle class mother will chide the child against 'soiling the hands'. But the natural instinct of all children is to play with sand and mud.

In the 27 volume book reports 'State of the Indian Farmer' (2004), Ministry of Agriculture which we have referred to often in the earlier book and this book), the Volume 1, is Overview, the second Volume 2, is on Land Resources. Without soil, there is no question of agriculture, farmer and other living beings. This Volume 2 is full of authentic data and summary of research findings about social, economic and technical aspects of land resources. Soil conservation is also addressed. The data is not merely on a national macro level but covers state wise data as well. The report points out the need for further desegregation. Some additional information is available in Volume 10 of the series, Rainfed Agriculture. In addition to quoting from these, we will also give some glimpses a few details of soil biology.

First and foremost, we need to understand the full set of dimensions about the soils (lands) in India as well as the dynamics of government policies, market economics and human actions/conditions on the soils. There are other elements of nature like water, rain, winds etc., are also affecting soils.

A quote from Volume 2 "In India, net addition to arable land became negligible and almost stopped in the early nineties.

In fact, going by available land use statistics, we discover that there are net outflows from the arable land in more recent years. Future land scarcity would be determined both by the magnitude and quality of land reserves. The nature of marginal land being brought under cultivation will also play a role in future in determining the quantum of fertilisers use and/or the pace and pattern of irrigation development".

The book Volume 2 also succinctly points out that our approach to soil/land cannot be merely scientocratic or technocratic from the view point of agriculture. A quote: "Land plays a dual role in rural India. Aside from its value as a productive factor, land ownership confers collateral in credit markets, security in the event of natural hazards or life, contingencies and social status."

Since we have many intellectuals in the country who depending on macro national data on a few parameters, jump to policy prescriptions, Volume 2 has a cautionary statements based on extensive research findings:"Interestingly a large share of area under non-agricultural use is not an important reason for high share of area under land not available for cultivation". The book further points out that "Maharashtra, Gujarat, Karnataka and West Bengal which have had high levels of urbanisation or industrialisation or both, have consistently had share of area not available for cultivation below the national average."Therefore those who attempt agriculture versus urbanisation or industrialisation have to be cautious and learn to look into details of locations and not generalise. This is borne out for other factors like soil productivity etc., as well, as can be seen elsewhere in Volume-2.

Another stark reality needs to be understood especially in the context of saving soils and eventually enriching them. To quote Volume 2, "................land-man ratio among sub-marginal holdings faced a continuous decline in most states of India, and in economic terms, the vast majority of tiny farm operators were obliged to seek their living from a strikingly low land base.

"The huge proportion of tiny holdings thus continues to be a major agrarian handicap. In other words, economic viability of an overwhelming majority of cultivating households, most markedly judged from on the basis of farm incomes alone, is a big question mark."

This agrarian handicap has become more severe during the end of the 1990's that is around the turn of the century. It continues through the 21st century though more than a decade has gone.

Soil conservation in India cannot be addressed without addressing the fate of these persons. Science, technology and extension work alone cannot solve the problem.

As for statistics, we are quoting the figures taken from Volume 10 which has a table (in p. 135 Table 5.3) and from Volume 2.

Total reporting area	305 mill. ha
Gross cropped area	185 mill. ha
Net sown area	142 mill. ha
Waste land (estimated)	64 mill. ha
Area under forest	68 mill. ha

Area under non-agricultural use was 9 million ha during 1950-51 and it grew rapidly to 20 million ha by 1990-91. But there was a steep decline in barren and uncultivable land from 38 to 21 million ha. Apparently, this is mainly due to the shift to non-agricultural use (which is a good sign). On the contrary land under cultivable waste field from 25 to 15 million hectares; this shift does not seem to be added to any productive category like pastures, area under tree crops or net sown area.

The Volume 10 also makes a statement that the net source area of 142 million ha "apparently is the limit of the extensive margin of land".

There is no point in crying over the spilt milk. One

could keep on pointing that "this could have been done" or "that" etc. Blames may be laid. But a realistic approach would be to conserve the land under net sown area now and make it sustain ably productive, and economically vibrant. And to reclaim much of the soils, wastelands, cultivable waste, pastures and fallow lands, etc., in a manner that they will be rich soils contributing to nature in terms of forests, tree covers, pastures etc., and also to maintain the biodiversity.

To achieve this, we require a combination of government policies to support the land owners, land-workers, technology-extension-workers, and entrepreneurs who will use land productively and sustainably, and some financial support as well as participation by communities, rural/urban people, intellectuals, media etc.,

Conserving soils to keep them productive and enriched (not exhausted) will be a crucial task in the 21st century in India and around the world. Youth would need to be enthused to take up careers/entrepreneurships in soil conservation activities (not as a sacrifice in their lives but to make a good living as well).

We will explore some elements in that direction in the subsequent sections of this article.

Soil Conservation

Let us start with a negative note from a quote in Volume 10. "Lamenting the absence of any policy intervention Nadkarni points out that soil conservation is not being adopted even where it is economically profitable for the farmer to do so. The degraded lands are 'Nobody's Child' neglected by the government as well as farmers owing to absence of arrangement to ensure that the benefits of investments made by the farmer accrue to him".

The Volume 10 (Page 136-139) further points out: "it is important to realise that the land use is influenced by a host of factors like comparative advantage in alternative uses, policy regime for agriculture and farmer's perception and attitudes".

It further goes on to describe about two major situations in India accounting for substantial part of land degradation problems:

(i) the green revolution areas and

(ii) semi-arid drought prone areas.

Citing example of Haryana for green revolution area, it is pointed out that the problem of salinity, unsustainable use of ground water, water logging etc., is "the outcome of the historical processes which began in the late sixties in response to the critical food situation faced by the country"............... "when the strategy devised to deal with a crisis was not suitably modified when crisis abated."

It is further emphasized that "the situation can now be dealt with only when thorough going reforms are implemented.

(a) to benefit from the export of wheat and rice, and

(b) to price power, water and credit to reflect their economic cost and

(c) to rationalise public distribution system and the related procurement, storage, and distribution of food gains. If the reforms are implemented, the problems given rise to by overexploitation of ground water will be eliminated at the roots. If the reforms are not implemented, the problems would persist irrespective of the attempts made to regulate, water use and to treat the degraded land."

Strong words, but it is the reality. Upper end agriculture requires large scale reforms giving economic freedom even while removing the populist sops given earlier. They will diversify for exports of high end grains. There will, of course, be fear of transition. But awareness and knowledge inputs to all the stake holders can be given to effect the transition.

But the policies in almost all states and the centre are going in the opposite direction! Let us remember that our precious soils cannot be taken for granted too long, with intermediate costly technical fixes!

Now coming to the **second major problem situation in the soils used for agriculture in semi-and drought prone regions,** these account for over a half of Indian agriculture.

"These regions are deteriorating into hard core poverty areas with low crop yields, stagnant technology, meagre investments and continuing degradation of land base owing to **erosion, loss of tree cover, and indifference towards conservation of scarce water resources"**

"Governments show little interest in these areas, farmers too weak to counter the oppressive poverty and rural communities subsist on periphery cut off from many dimensions of modernisation".

This is not the type of living with nature we should look for or even allow to happen to our fellow Indians.

Volume 10, again points out, that technologies for overcoming degradation of land and loss of water resources are known. In addition there are technologies waiting in laboratories which could substantially raise crop yields.

"There are instances, regrettably few of village communities achieving remarkable growth and modernisation through local leadership".

We have witnessed such a modernisation in some parts of Bihar, UP, Uttarakhand etc., under the leadership late Prof. S.K. Sinha in the Vision 2020 follow up programme of TIFAC.

Sustaining Successes to a Movement

All such successes keep the hope up. But when will we cover millions of hectares of such agricultural lands and also other lands (pointed out earlier in the statistics) whose soils have to be reclaimed, rejuvenated and conserved?

In areas where there are good governmental programmes and implemented well such as in some of the integrated water shed development projects for rain fed agricultural areas, long term sustainability is poor. After initial efficient management, there is a slow degeneration to old ways.

Quoting some of the research findings, Volume 10 concludes that "one of the reasons for this is the extra care taken by the implementing agencies (mainly government agencies which provided all the technical and engineering help along with subsidies) and managing the projects from the above. In the process the participants became mere onlookers than stakeholders. Even the efforts to form collective associations and operationalise them have met with poor response". I recall the late Prof. S.K. Sinha's participatory efforts in Bihar. He did not start till the farmers asked for, though he was spending lots of time over a year to give them information and options and to explain to them. He made them do it, they invested. He brought the best know how, and did soil testing etc., at TIFAC's expense. He nudged them towards being market savvy and encouraged them to form associations. They did of their own. The other successful efforts by others would have followed such approach.

Much of the government interventions had been experimental covering only 2 or 3 million hectares over a decade or so.

In addition to the above aspect of farmer/stakeholder participation, **"the influence of the larger agricultural system needs to be borne in mind while looking at**

the cumulative neglect of land and water resources in the drought prone-area”. The Volume 10 (page 140) further states” the market environment marked by low and unstable prices adds to the woes of the farmers facing droughts and scarcities.” **“The policy regime (by governments) instead of providing a corrective makes the situation worse by neglecting infrastructure, investments, and institutions”.**

In addition, we should also be aware of other constraints to the farmer: **“There are several other constraints at the micro level that may not allow farmers to go for the most efficient crop even when price incentives are given.** These constraints range from limited irrigation to lack of marketing or processing facilities or even lack of farmers’ knowledge about the crop......” These observations were made based on the research on oilseed like diversification.

To Sum Up Actions Towards a Movement for Soils

For protecting, rejuvenating and sustaining the land/ soils under agriculture we require two broad classifications as explained earlier.

(1) One for green revolution type of well irrigated areas
(2) Another for a large segment of poor farmers in semi-arid areas.

For the former, various forms of economic reforms are required along with preparing them for the transition. Even here what is good for Haryana or western UP may not fit Andhra or Punjab. Sufficient amount of localisation is required with flexibility to provide the needed transition.

For the second category actions are more complex. First and foremost we should remember the statement in Volume 2 (p. 252) **“It needs to be understood that land degradation is a localised problem and the nature of activities leading to it is even more so. Therefore there needs**

to be micro-level solution of such problems".

While making a large scale mobilisation of farmers, people and government/NGO machinery for all the semi-arid region agricultural lands, with the above in mind, the projects (Movement" will be a better term) should take care of what is earlier elaborated.

(i) farmer/stakeholder participation

(ii) understanding of the impact of the larger agricultural system

(iii) Adequate provision of infrastructure, investments and institutions by government partly and through enabling policies for ownership of assets etc.,

(iv) Other micro level constraints in each of the localities.

Technologies are known. Let them not be "pushed". Also there is a need to enable/deploy private sector companies to provide these with the above precautions. Without being needlessly intrusive or control oriented, the policies for each of the local regions may allow for farmer participation in these companies as joint partners or leasers of lands etc. It is also necessary for Government to enable to develop a number of micro-small-and medium enterprises (MSME'S) around many of the villages to generate more wealth for the people/families around that areas.

This will help them to "feel better" in terms of financial security, through employment (part or full) in these MSME's. Better incomes will also help them to invest on their lands to improve its quality's or to contribute to "common goods" like pastures, forests, tree cover etc., around them.

These MSME's may be established on waste lands or non-cultivable lands around those villages, so that the better soils are protected for agriculture, pastures, forests water resources etc., This helps the nature to regenerate, rejuvenate and enrich the local ecology.

Some of the wastelands and non-cultivable lands could be used for services sectors – for the supply-chain-management-networks required for inputs to agriculture, animals, birds, fisheries etc., and also for the outputs of marketing services such as storage, cold chain etc. Private sector companies or cooperatives (rural and urban) may take these up as commercial venture if policy is conducive. Only when the economic activities are large and income giving, people will take interest in conserving the precious resource: SOILS and related water resources.

Some of the wastelands could be used for locating waste-water-recycling and reuse plants. Again such decisions should be local, instead of being taken in New Delhi or even State capitals. Only with such an innovative approach to governance (which can blend high science knowledge inputs, business interests and individual family interests and people participation), can save the precious soil and associated water resources in India. One point slogans (such as Industry versus agriculture, don't-touch-agricultural lands, etc.), do not save soils nor give the farmer's livelihood nor lead to economic growth. Farmers will be pushed further into poverty. Soils will degrade without attention and resources.

Prosperity of the farmers cannot be merely on land alone as we have seen the statistics (land-to-man ratio, nature of lands). It requires a multipronged approach to save the soils of various types, and use wisely some of the non-cultivable and other lands for industrial/economic social/entertainment activities. The saving of soils and keeping them rich and active is not a one-time action but a regular activity.

We should also from now onwards, shun various other products which had been developed over millennia by using the top soil (which is really rich) such as bricks, potteries etc. The burnt soil cannot be easily recovered or recycled. The temporary "green" middle-class euphoria over the use of "Kulhad" cups in railways and hailing it as the return to Nature, is a danger to soils.

Let us use alternate means for such needs of bricks, pots etc. The population has gone up very high and consumption is much higher.

Top soil is precious; the whole of soil from top to bottom is precious. **NOT for minerals or materials but in itself. Its biology sustains LIFE in nature.**

Soil Biology

In India soil is mostly studied as classification for agriculture, forests wastelands, etc. Many of them are macro or at best, meso classifications. Very little studies are there in terms of micro level status of various soil holdings. Civil engineers study the soil in detail before construction. Such tests and classifications are a part of the engineering course work. But they do so only from the viewpoint of safe and efficient construction of building, roads, bridges etc.

The curriculum of agricultural scientists contain considerable details about various types of soils from agronomic point of view. Very little is studied about soil biology. Mention of earth worms, vermiculture, or compost manure alone do not form detailed study of soil and its biology. It is a relatively new discipline in terms of interest of mainstream scientists since 1990's. The technologies we have referred to earlier, as available for soil conservation and preservation, are mostly derived by practical and empirical knowledge of trials in many fields. They are useful, practical and applicable immediately.

So also the experience of organic farming community which is currently much more an art of application, is useful.

Real scientific insights to preserve soils in its totality will come during the coming years and decades with researches and deeper understanding of soil biology and ecology. There are already a number of standard text books on these subjects including with the titles soil micro biology. Some of them are down loadable free. We give some extracts from a Wikipedia on "Soil Biology".

Some Quotes

"Soil biology is the study of microbial and faunal activity and ecology in soil. Soil life, soil biota or edaphon is a collective term for all the organisms having within the soil. These organisms include earthworms, nematodes, protozoa, fungi, bacteria and different anthropodes. Soil biology plays a vital role in determining many soil characteristics yet, being a relatively new science, much remains unknown about soil biology and about how the nature of soil is affected".

When we talk of biodiversity it is mostly understood to be above the soil; forests, lakes, coastal areas, their flora and fauna etc., are thought of. Some awareness, in India, has come about biodiversity in the oceans. Still much more sharply about the commercialisable/useful products from the seas and oceans. But a large proportion of nature's biodiversity is under the soil. Also, little is known about the soil organisms and soil functions; scientists have observed that linkages between them are incredibly complex. The soil "food web" is interconnected. We need to understand these interactions with living communities that exist within the soil.

As of now, most of us just know that soil breaks down organic matter and makes available nutrients for plants. Very few of us would know that these soil organisms also store the nutrients thus preventing their loss by leaching. We are aware that earth worms are important for bio turbation in the soil but know little that microbial exudates act to maintain soil structure.

To use the actual quote from the Wikipedia "………we find that we don't understand critical aspects about how these populations function and interact. The discovery of glomalin in 1995 indicates that we lack the knowledge to correctly answer some of the most important question about the biogeochemical cycle in soils. We have much work ahead to gain a better understanding of how soil biological components impact us and the biosphere".

This is an area with fresh new opportunities for scientific research: Research that is going to be immensely useful for saving LIFE on earth and for living well as a part of NATURE. During the 1960's J B S Haldane, a great versatile British scientist who made India, his home used to plead with Indian scientists to do research on outdoor biology, as India was rich in biodiversity. It would have created some great original scientific schools in India, with many new discoveries attributed to Indian scientists. But the managers/leaders of Indian scientific and academic community chose to be predominantly – follower researchers, that is, following what is done in the English speaking West.

My appeal to those associated with 35,000 odd colleges in India is to attempt in some direction to learn about soil biology and start researches. Many of the courses containing microbiology, home science, biology etc., may also begin such a process. Rewards may be high. **Of course, I will be delighted if the elite science institutions and agricultural Universities start active research in this area, in different parts of India.**

I would also appeal to our scientists to get a few of the standard new text books on soil biology/ecology/microbiology translated in Indian languages so that many more Indians can read and understand. Just as Space and Oceans are important to **study and explore, exploration and study of the top layers of the soil and learn/understand what is happening in that huge "world" is much more important. Our entire food chain depends on that world!**

Let us have a few glimpses of soil biology from the extracts of the Wikipedia.

"Soil biota includes:

- Mega fauna: Size range 20 mm upwards e.g., moles, rabbits and rodents.
- Macro fauna: Size 2 to 20 mm e.g., wood lice, earthworms, beetles, centipedes, slugs, snails, ants and harvestmen
- Meso fauna: Size 100 micrometers to 2 mm e.g. tardigrades, mites and springtails.
- Micro fauna and micro flora: Size range 1 to 100 micrometers e.g., yeasts, bacteria (commonly action bacteria) fungi, protozoa, roundworms, and rotifers.

"Of these, bacteria and fungi play key roles in maintaining a healthy soil.

They act as decomposers which breakdown organic materials to produce detritus and other breakdown products. Soil detrivores, like earthworms, ingest detritus and decompose it. Saprotrophs, well represented by fungi and bacteria, extract soluble nutrients from delitio. The ants (macro faunas) help by breaking down in the same way but they also provide the motion part as they move in their armies. Also the rodents, wood eaters help the soil to be more absorbent".

Even this brief account gives only a peep into the complex interconnections between the different "lives" (organisms) under the soil. Excessive chemicalisation can seriously disturb the dynamic ecological equilibrium within the soil; as also excessive watering or excessive drought.

Wikipedia gives a fairly detailed Soil life table.

Having had a glimpse of the above, let us understand the scope of activities under the broad title soil biology.

- Modelling of biological processes and population dynamics.
- Soil biology, physics and chemistry: Occurrence of physicochemical parameters and surface properties on biological processes and population behaviour.

- Population biology and molecular ecology: methodological development and contribution to study microbial and faunal populations: Diversity, and population dynamics; genetic transfers, influence of environmental factors.
- Community ecology and functioning processes: interactions between organisms and mineral or organic compounds, involvement of such interactions in soil pathogenicity; transformation of mineral and organic compounds, cycling of elements; soil structuration".

It is further added that complementary disciplinary approaches are necessarily utilised which involve molecular biology, genetics, eco physiology, biogeography, ecology, soil processes, organic matter, nutrient dynamics and landscape ecology. Imagine the associated measurements and instrumentation. Not to mention computational inputs which would involve complex simulations.

This is the type of NEW AREA OF SCIENCE which is awaiting our scientists, academics and the youth. Are they ready?

Before ending, I would like to give some numbers. We are used to newspaper reports about thousands of billions stars in a galaxy, billion and billion galaxies in the Universe. Look at some numbers for life under the soil:

- 100 million to 3 billion bacteria in a gram of soil.
- One bacterium, through multiplication is capable of producing 16 million more in 24 hours.

Hostile conditions will not kill the bacteria altogether; rather they will stop growing and wait for better conditions. This is what gives hope for reclaiming the soils. Bacteria are also colonised by persistent viral agents (bacteriophages).

- A gram of garden soil contains about 1 million fungi such as yeasts and moulds ....you may read more on Soil Biology from various resources.

I should also mention that biology of roots of plants and trees is another large area of study and research.

This will also become a sub-set of soil biology. We are not even giving any glimpse of these.

Let me again appeal to our academics to get interested in soil biology. From the listing above, there is a plenty of scope to accommodate many super-specialised disciplines.

Conclusion

Soils are precious resources. In India, most of them are in danger. They need to be reclaimed, rejuvenated, conserved and enriched. We need to do so, on the lines suggested in this chapter with known technologies and sciences. Many other parameters like economics, farmer's aspirations, markets etc., are also involved, as explained herein.

THAT IS THE URGENT TASK TO BE DONE AS A MOVEMENT. Let us not wait for any new ideas or inputs to come. It is more critical for India than climate change issues, in the short and medium term.

The long term will also need a deep understanding of soil biology. Let us create interest in it right away. For the new generation, let it start at school at pre-primary itself and primary and above. Not as an additional subject to "rote" lessons. Let our children play with mud, soils in the garden and field. Look at worms, ants etc. At higher classes they can see the microbiological creatures in microscopes and also study about interconnections by actual "touch and feel" experiments.

"Soil is the Essence of Mother Earth and Nature"

Let us preserve it.

❐

8

Chaotic Churning: What is Ahead?

The unconventional and straight speaking John Kenneth Galbraith, former US Ambassador to India, described India as a "functioning anarchy."He probably did it fondly. That India which he described was over four decades ago. India since then has passed through several stages including a period of emergency and several coalitions which lasted for brief periods. It is amidst these processes, India quietly jettisoned its autarkic socialist economic system to follow market-economics (1991 liberalisation). This was done amidst deep financial crisis by a minority government. By then the global changes were drastic. India's traditional friend, the Soviet Union had collapsed. The neighbouring Afghanistan was falling into the hands of religious extremists. Unnoticed China was growing very fast, after embracing market economics in 1980.

Looking back, what happened during the two decades of liberalisation and globalisation, which also coincided with ICT revolution world- wide and in India, it appears to be a dream world. Suddenly the consumer-goods-starved India was flooded with many consumer goods from all over the world! A country which waited for 3 to 4 years after booking to get a Bajaj scooter or an Ambassador car, was flooded with many automobile assembled/part-made-in-India! So was with the explosion of TV channels. Mind boggling expansion of professional colleges. No wait time for telephones or cars or LPG cylinders.

For the first time Indians started seeing so many wide roads, linking India as a golden quadrilateral. Textile boom made almost all Indians wearing better clothes. No wonder, slogans like 'India Shining' started appearing during 2003.

Not All Shining!

Amidst these developments, India weathered many internal turmoils. Cross-border-terrorism was on the rise; with the Kargil War, it peaked. The frenzy of demolition of Babri Masjid (1992) leading to Bombay blasts (1993), also increase of Naxalite/ Maoist violence. A number of poor people also were lifted to lower middle class levels; aspirations of the middle classes soared very high; but severe poverty remained especially because of the liberalising powers-that-be-were captured by the lure of IT and totally ignored agriculture and manufacturing. They dreamt of India's leap frog to Services economy. Even for serious institutional and social cancer like corruption which was also growing leaps and bounds, their solutions was e-governance.

The 'anarchy', within Indian institutions, social and economic systems were hidden from the public memory, and the reports of the analysts/experts, and the media hypes! Sensex became the new watch word for national performance as it was with cricket matches.

Many NRI's (Non-Resident Indians) flew all over India to give various forms of advice to make India a super-power, and to usher into "knowledge economy". Many spiritual gurus emerged as well in the economic exuberance of the rich and middle class.

Black money 'freely' flowing into (and generated continuously) the Indian economy, was a great lubrication to keep the growth machinery moving. It partly redistributed the ill-gotten money in expensive social shows like opulent marriages, great functions of leaders and award ceremonies, fuelling electoral machineries and "purchasing the powers-that-be" to have privileged access for themselves. The "connected" became more "connected" through Indian black money that was available globally.

As per the global Financial Integrity (GFI reports (Times of India 13-12-2013) in the decade 2002-2011, India lost a staggering Rs. 12.7 lakh crores (344 billion) worth of back money created through various illicit and criminal ways as outflow. This worked out to an average of Rs. 1.6 lakhs crore outflow for every year of the decade (about Rs. 1333/- per Indian per year). During 2011 over Rs. 4 lakh crore worth of black money was Illegally taken out of India, as per a new report of GFI (that is about Rs. 3333/- per Indian per year) This report is based on tracking global money transactions. What is done purely as cash as in drug and human trafficking etc., is much more!

What a colossal waste of wealth! Had it been within India and flowing even illegally as opulent expenditure, a number of poor/middle class Indians would have benefited and in the jobs so created!

Can we call such colossal creation of black money and its outflow as a 'functioning anarchy' or 'organised kleptocracy'? Is the Indian people's *khazaana* infinitely deep?

Media and activists only shout about the corruption and scams. It is being narrated as "thriller stories". Root causes are not addressed. But to get political gains, government started wasting precious tax money (only a miniscule percentage of India pay income tax) inpopulist schemes such as giving free PC's to class 12 students; free colour TV's; now even daily cheap food cooked/served by government, free wages as an unemployment guarantee, money for marriage of girl, etc., in addition to various subsidies for LPG, petrol, diesel, electricity, water etc., are they meant to reduce the perceived "anarchic anger" of the people against the politicians and bureaucrats? Or to buy their votes to come to power again? How long can these sustain?

Slow Down and Implosion

The economic growth has plummeted because of several such huge misdeeds. There is loss of jobs of many persons who were in reasonable jobs before. New jobs are hard to create. Populist doles cannot fulfil what can be achieved through self-respecting hard work and good incomes earned thereof, thus adding to the self-esteem of people.

So India appears seriously chaotic. Supreme constitutional bodies like Parliament (Lok Sabha and Rajyasabha) and State Legislative assemblies are creating scenes worse than street fights. Most constitutional niceties have been dumped. Some constitutional authorities not only do not practise the institutional niceties but also transgress on "turfs" of others. Courts are already crushed under the weight of litigations but also have to enter into the cases filed through such constitutional transgressions. Allegations have become common place, filling media space. Most TV channels do not appear to be news channels but look like street fights in the name of panels with the anchors going frenzy and appealing to the 1.2 billion of people of India to judge as if all of them understand English or Hindi.

In the process, the Indian elite society at least, seems to be imploding with its own one-point agendas. Indian government appears paralysed to take any economic or foreign policy initiatives because one or other state government opposes something or another. Also most of what is happening around the world appears to be irrelevant for Indian elite: Middle-East, Syria, Ukraine, Nigeria, Central African Republic, Myanmar, Nepal, Bangladesh, Venezuela, Iran, Saudi Arabia, UAE, Europe, Japan etc.,

Media and middle class spend more time on the Italian Marine case or the case of an Indian diplomat! Earlier we had seen the jingoistic campaign against Norway, over the custody of two children, as per their laws. The case ended up as a comic tragedy! Before that there was a "war against" racism with

Australia! Now we have unabated crimes against women, men/ boys of different looking ethnic origins etc., in the national capital!

Almost in any place in India, traffic and normal life can be stopped by any group including by the parties which rule. Chaos in the national capital was heralded by some as a saviour of India, emerging with alternate politics. The breakdown of day-to-day law-and order situation coupled with "free movement" of various terrorists who can strike at will, does not augur well to attract investors (apart from chaotic and sometime draconian tax administration).

Let us look at a recent report (23 Feb 2014 - Times of India): There is only 1 police person for 1000 Indians. On the other hand, there is 1 NGO for 600 Indians. Since a number of states have not given report to CBI, it is possible that there is 1 NGO for 500 Indians. That is, 2 NGO's per 1 police person. Is it a positive sign of cooperative social regeneration in India? Or is it a parasitical phenomenon for using loop holes in various tax laws and also as a base for various forms of activism? How many of them are conduits for flow of money from foreign entities? Unfortunately researched and authenticated data on these will be hard to come by. There will of course be a few very good NGO's.

But the ratio of police personnel for number of citizen is very poor; recommended figure by UN is 1 for 222 people. However a question is whether just a mere increase of police would improve the law-and-order situation? Will women be safer? Will crimes come down? Or will it increase police Raj? Or increase the number of cases in the courts, only to languish there, with many suspects being under judicial custody for years without hearing?

Unfortunately, we need to list many questions because in the "churning" that is taking place chaotically, we are unable to get clear answers. Also Indian social science researchers seem to keep away from studying real Indian issues scientifically. Therefore, we all form opinions and not statistically derived clues and causal connections.

What about the Bulky Base?

I have a deep suspicion that most of the "chaotic churning, we see around us, the litigations, PIL's allegations, protests, dharnas etc., are limited to the elite class and middle class, occasionally drawing the lower middle class and poor persons for big morchas or as audiences for the big election campaigns. The poor in the bulky base and lower middle class persons, even those who are recipients of populist doles really know in their hearts, that they cannot take solace in these chaotic churning processes. They are born with burdens of poor and hard lives, in their families and during their childhood. Despite decades of slogans, promises and propaganda, actual improvement in their lives comes about extremely slowly - be it for economic upgrading or social improvement. They may see the TV fights of political and elite class, also shown in many Indian language channels, but may be treating them as a movie show!

When I go to the interiors of India or Tier 2, tier 3 towns and talk to the people there, their interest in the so called big 'emotional' issues stoked by the state leaders or shown in TV media, is practically negligible. They may have their local politics to "chew upon" when they gossip. Perhaps they may be stirring up only during the 2 to 3 months of major election campaigns to discuss some issues. Even then, it is unlikely that they appreciate what goes on in the national channels all 365 days or the "chaotic churnings" in New Delhi or Hyderabad or somewhere else. Even fears from a terrorist attack or major rail accident or communal clashes, may not be bothering the bulk of them because they have to earn their daily living. Most of them have learnt that the divisive or vocal politics or big noises about Bills, make hardly any difference for them.

This is a sense which I get when I visit or by talking to people who go to interiors of India. In one sense, it is good news because the cancer (of depressive pessimism, and intolerant passions and a craze for domination, which are present in the dialogues of the elites and the media) does not appear to have spread to them.

Agitations in Urban Centres

Another empirical fact that supports this hypothesis is the fact none of the "protest attempts" have evolved into bigger movements despite repeated show of scenes and discussions held in the TV media. Anna Hazare's fasting did create some swelling of crowds, but they were mostly from middle class form around Delhi area and some might have been transported from other metros. Again the four day chaos and din at Delhi before the Republic Day, fizzled out.

At best there can be one or two days of "successful" bandhs in cities because people fear violence to their vehicles or buses. It is all over and back to normal after that. Rural India and smaller towns are unaffected by it.

Most of the "agitation" leaders are either forgotten or "absorbed" into the "chaotic system". In that sense it is a "functioning chaos". But the damage to the governance system due to these chaotic and unpredictable agitations is high, though most of the chaos are really localised in the national capital or in some specific localities as it was in Koodankulam. Even for the tragic rape-murder case in Delhi, which resulted in spontaneous agitations, it was mostly confined to Delhi. An ordinance was rushed through. But many tragic and atrocious rape-murders continue to take place in Delhi and other cities of India (we do not get to know the cases from villages and town, as they are unreported or get buried in local language press). I wonder what happened to the swelling crowds. There was none even for the atrocious killing of an Arunachal boy. The tragic episode has been buried in TV shows on racism and demand for anti-

Racism bill. One demand during the Delhi rape case, was instant justice (which actually meant for most of those who demanded it, instant hanging of the rape-accused), and it is yet to take place. The cases are with courts. So also for many other rape-murder cases which took place later; in some cases even the culprits have not been booked.

Similarly even after the swell of Anna Hazare fasting, later passing of the Lok Pal Bill and acorruption-agenda based party coming to "power" in Delhi State, the life of ordinary middle class person has not changed, petty corruptions for day-to-day lives still continue. Also those who agitated against corruption continue to pay such corruption, to get things done.

Thus the hard realities of India – the hard and uncertain lives of the persons in the Bulky Base and the lower middle class Indians, as well as the basic economic insecurity of the middle persons (which will prevent them from long drawn agitational paths) – are keeping India "stable" despite many "chaotic churnings" that take place in the political and other elite circles. Only time I had seen a large "people-student" movement in most parts of India, was during the JP (Jaya Prakash Narayan) movement during the early 1970's. It was more in the North and West than the South and east. Even in Jammu and Kashmir, protest movements and stone throwing fizzle out after a few days.

Seeing such empirical information, we can conclude that the "chaotic churning" being all around is unlikely to turn into mass movement or a mass anarchy or mob violence, as it appears to be in Bangkok, Ukraine, Bangladesh, Egypt etc., Perhaps some external forces are exploiting the deep divisions within these countries and actively supporting their "crash" into "chaotic conditions". To that extent, India can be considered a much stable democracy.

A Glimpse of Other Countries

Functional stability for countries like Egypt, Pakistan,

Thailand etc., not to mention those which are under an autocratic rulers, are provided by the armed forces. For the fledgling and fragile democracies be it Bangladesh, Nepal, Ukraine etc., the role of the developed nations (west) is questionable. Human rights or some other issues of recent past, are invoked to create for its rulers, some serious divisive troubles.

While liberal ethos and democratic values of these western countries within their own national territories have to be appreciated, their own record of human rights or any other liberal values (which they vouch as their highest principles) vis-à-vis peoples of other countries is miserable and contemptible.

About five decades ago the poisonous herbicidal attack, USA did on Vietnam is to be seen in numbers: 34 C-123 huge transport aircraft, sprayed 75 million litres of herbicides including 40-million litres of dioxin on a small country to destroy its virgin forests in which Vietnamese guerrillas operated. There was so much of diotoxin in those 34 aircrafts, (which were later used by US Air force) that their own personnel were affected. It is well known and documented. The Iraq and Afghan wars, and the support to Syrian rebels, have serious violations of lives of civilians, still no leader or administrator from USA will be prosecuted for human right violations, because it is done in the USA's national interest. Let us not forget that Taliban was created and supported by USA to fight the Soviet Union in Afghanistan.

But they prosecute "as per their law" leaders of other countries; denying them visas; or issuing court orders on them to appear in a US Court. They move the UN Court. They move UN against the Heads of states of other developing countries who have protected their democracies and governance systems against severe and savage terrorist and secessionist forces.

So for USA, the external world is to be subservient to its national interest. It has the Right to invade the privacy, Human Rights of other countries, even its elected leaders, even from its ally Europe. Other European countries have similar double standards. They begin a huge propaganda war against other countries for relatively petty human right charges. Some of their rating agencies are pressed into service for bringing out figures: India is rated very low (near the bottom) in freedom of press.

There is no point in being moralistic about it. We need not have to take high moral grounds and preach to the world as we did during the 1950's and 1960's. What is respected world-wide is the strength of the economy, the Agility of the Security Systems (includes military, para- military, police etc.,) and the unity of the people.

In the current world of global communications, the new world war (which I believe has begun with so many flash points of country wide "civil society" agitations in many crucial countries encircling Russia, China and perhaps even India, Brazil and few other emerging economies) has begun for world domination. This war is "the war to be fought in the minds of men/women and even children" to modify the famous Churchillian phrase on future wars. That is why we as people and country have to be aware of these external developments and not get lost in the impressive "chaotic churning".

What is Ahead?

Even during "normal" times, future is very difficult to predict. While writing this piece, I saw a quote from Vladimir Ilyich Lenin "There are decades where nothing happens and there are weeks where decades happen".

During the first decade after India's independence, there was more euphoria of great freedom even while dealing with devastating human tragedies of partition of the country;the subsequent decade saw the establishment of many new institutions like IIT's, ISRO, and expansion of other Science and technology (S&T) infrastructure. There was beginning of Green revolution and setting up of many major public sector industry. During this period India was also jolted to the realities of geopolitics through a humiliating defeat at the hands of China.

The subsequent decade (1967-1977) saw tumultuous domestic politics, emergency etc. But geopolitically there was euphoria with the creation of Bangladesh (partition of Pakistan) with the active assistance by India. Economic growth was slow due to the commanding heights of license – permit – quota – inspector - raj and heavy tax regimes. The seeds sown and saplings grown of corruption and crime in public life due to politics – bureaucracy – crime - business nexus started growing as big thorny trees affecting day-to-day lives of people, the middle class which had grown to some size by then started grumbling, sometimes agitating. Workers in the organised sector (who were really the labour aristocracy in an India where most of the labour was in unorganised sector) showed their power through the strikes and gheroas called by the unions. It was also depicted as the dream of the socialist India. The subsequent decade slid through with feeble attempts to open up Indian economy, that period also saw major scams and loss of faith in the politicians – bureaucrats. Things drifted with chaotic changes in governments.

Then came the liberalisation of 1991 and "IT revolution" about one and half decades of rapid growth, and resultant consumerism, grew the middle class to a size much bigger than many countries in the world. It was a great attraction for global investors/companies and also for the Indian private sector.

The rapid economic growth created more wealth, therefore more taxes. The coalition at the centre and its advisers during the past decade as well as State level leaders felt that the collected tax money can be given as doles to the still existing huge Bulky Base (BB). These persons at BB can be given an impression that it is the leaders themselves who are giving their personal wealth. Then the political leaders and their advisers can perpetuate as rulers. But the process created inflation, following the simple laws of economics: "free" money flowing into the economy without creating some products, creates however demand. The demand exceeds supply rapidly, thus prices rise! Also the tax money spent on unproductive, non-asset crating tasks does not add out economic growth, nor becomes the reproductive wealth"(Please recall the Modern Mantra given by late Shri C. Subramaniam, as we had pointed out earlier several times).

The middle class is hit with a double whammy: slow down of economy and galloping inflation. Many of those who are in job now do not get better increments and new aspirants do not have job openings. Thus this class which had tolerated corruption and the hassles of Indian Government rules/laws is no longer able to tolerate the mega scams associated with their political leaders, and their patrons in the private sector (crony capitalism is the new word getting popular, though during the licence-permit-quota-inspector raj decades, it is only the crony capitalist who could get the licence easily and be able to "manage" the oppressive tax/approval/inspection environment).

For the upper middle class, the slow down and inflation affected their newly earned lifestyles. They had to cut down on many 'goodies', they were getting used to, they felt seriously deprived. Auto loans etc., started getting costlier.

For the investors, the "erratic" behaviour of tax officials (as the investor perceives) added a fuel to the already existing hassles. So they quietly move away from India, adding further problems for a recovery from the slow down.

So the future suddenly started looking bleak for many of upper, middle and lower middle class persons. What is ahead of us? They need to chase some devils always! ! Thus there are many sporadic outbursts and some local violence.

But as we have argued in the earlier parts of this chapter, such an attitude of total pessimism has not yet come to them. They have a hope in the "fundamentals" of India, in whatever vague terms, they imagine it to be. That is the reason that they are in search of a "leader".

Most of them do not understand nor care to learn about the niceties of constitutional government. That is worrisome because such an attitude by large sections of privileged classes can lead to "loss of democracy".

But in the name of democracy, if a few "activists" or ambitions "opinion leaders" try to spread "chaotic churning" with tall promises of "total instant clean up" as well as immediate relief to the pressures of life through many unsound measures such as slashing tariffs, increasing subsidies etc., there is a possibility that such "leaders" may get some sizeable votes and seats in the Parliament. They can then continue to degrade the parliamentary institutions to further abysmal levels. Problems will not be solved, but the frustrations of the people will be stoked further.

If the votes are divided between many parties with many aspirant "Prime Ministers", it could be another form of chaos in governance, perhaps worse than any coalition, India had witnessed before, including the present ruing ones. If on the contrary, if these leaders settle down and bury their ambitions for a common cause of lifting the economy from the current morass, thus creating a feel good factor amongst all classes: Rich & Super Rich; Upper, Middle and Lower Middle classes and the Bulky Base (without doing divisive politics), then the India can get into a better trajectory with the "Chaotic Churning" being converted as a dynamic momentum for all Indians.

This can also happen if a single party or a single pre-electoral formation (preferably than post-electoral combination) get a comfortable majority; perhaps then India has greater chances of moving forward and upward in the economic and social uplift of its people. Again it depends on the leadership qualities of the persons who will form a collective team.

Therefore a lot depends upon what happens in the next few weeks (written before elections). If things turn out in a way to enable formation of cohesive teams with strong commitment to curb the chaos through positive and productive actions, which will engage all Indians towards self-respect-enhancing economic activities, then India's next few decades would have been shaped, in those few weeks.

I don't want to spell out again as to what will happen, if such a cohesive situation does not arise, as enough has been said in the earlier paragraphs about the dangers of continuing chaos and especially the continued loss of hope in future by the middle class, rich as well as the bulky base.

For reasons which have been given before, the 'chaotic churning' is yet only temporary in India; it has not yet settled as a societal attitude and feeling of hopelessness. If a strong leadership teams emerges post-election, soon India will transition to a 'functioning anarchy' first (during the transitionary clean-up process) and then towards a stable and disciplined India in which all Indians can realise their full potential and enjoy a good life.

May the weeks of April-May 2014, be another positive turning point for Indian people and India's history. I leave it to the readers to judge the path of India.

❒

9
Coping with Complexity: Macro Versus Micro

When this article (written before elections for Kisan World) appeared, India had gone through a major churning of macro – to – meso-to-micro, as described in the last chapter. It is not that the macro level debates, analyses, speeches, suggestive and non-suggestive imageries in the TV channels, advertisement posters, cartoons etc., drive the meso, and in turn the micro. Real life is much more complex. We have this as a chapter because it addresses a crucial factor which is eternal.

There is no clear method of understanding how an individual voter thinks at the time of voting or even when deciding not to vote. This micro level decision is not a standalone, free will decision. Many "forces" work on the voter: some from a very distant past; some from the near present, even a day or a few hours, before he/she votes.

Statistics-experts tend to decode these individual decisions by earlier trends; sample voter surveys etc., There are a number of micro-groupings which work – ranging from local issues of sanitations, or water supply or joblessness or promises by the electioneering politicians or caste/religion-considerations etc. If one tries to aggregate them, for one parliament constituency with an average of 16 lakhs voters, the interactions will run to trillions. Link these to the national level with so many candidates and the successful ones of 543 MP's. Then there are other complex manoeuvrings for alliances, cabinet making etc. Here again everything is not "macro" such as the policies of parties; in fact micro-level prejudices play sizeable roles as well.

The macro analyses are done to keep the readers of newspapers or viewers of TV channels engrossed. In fact, these are service sectors industries themselves. Most of these analyses fall flat within days and weeks.

When government is formed, all these channels, media, etc., will start discussions on a series of policy measures ranging from agriculture to manufacturing to exports to employment generation in heated debates. Some policies start working, but to claim victories of policies within 100 days is really a bold claim. But, of course, all these "macro" talks/policies can create a feel good factor in the vocal middle class space. How the minds of the micro level Indians work or feel are difficult to assess authentically. Sample surveys are nowhere near the near the real micro level details.

When the prices of daily-use goods like vegetables, food, milk etc., come down, people generally feel happy. If huge construction projects, road making etc., get energised, one can obviously "see" "employment generation". But the experts will question these "subjective" findings and will say that we have to wait for empirical data, which may take many months.

Thus there is a big hiatus between what we feel individually or in small collective groups and what are described as "authentic" reports by government, corporates, academics etc.,

These "findings" and "perceptions" sometimes lead to difficult, if not catastrophic conditions for many poor or marginally living people. Often macro policies are made on such findings and perceptions. Agricultural workers and small agricultural land holders (who naturally will have only a few birds or animals to supplement their incomes) are often the worst sufferers of such policies, as it is happening in India. And also many millions of workers (casual or contract) depend for their livelihood on micro and small industries.

We therefore desired to look at the theories of complexities as in economics and other systems like societies and governance. Thus I revisited reading a seminal book "The Origin of Wealth" by Eric D. Beinhocker (2006). The book's subtitle is Evolution, Complexity and the Radical remaking of Economics. Also an article "Are economic systems like Organisms" by Mae-Wan Ho, Institute of Science in Society (2010).

Biology, Evolution and Society

The mindsets of most of the educated persons are still conditioned by the deterministic theories of physics, chemistry and beauty of simple mathematical models to which all the Nature's Laws are reduced to. This is commonly known as scientific thinking. Most of us take it to be the finality of Nature, the ultimate truth. The philosophical implications of the early 20th Century physics and subsequent applications of its methodology to other fields like chemistry and engineering (especially in electronics) have not filtered through to most of the non-scientists. **Even many scientists prefer to be scientific workers in their own super-specialised fields and keep publishing papers, rather than worry about the issues of ordinary individuals. So will be engineers, doctors, administrators etc.**

They would rather leave these items to the economist who 'discover' many laws, use them to predict and thus give policy perceptions to government, corporates, banks etc., The assumption is that 'they' will take care, with their "scientific approach" and huge mathematical models requiring huge computers.

Are we safe with these assumptions of reposing faith on such "economic models" which also have disciplines of micro and macro and even econometrics? The book I had referred to gives a number of clues. In the past two to three decades, the world economies have gone through many crises. Many people in societies like India have suffered very badly and many are still suffering. But we should also recognise the fact that poor people (a billion or more globally and about 200 or so millions in India) have been lifted to a good degree of prosperity and to secure income levels.

Those from the fields of physical and engineering sciences give many "solutions" to pull the people out of poverty and misery - new seeds new products, new processes etc. Also many corporations come up to mass produce those products. In fact many totally unforeseen products and services like cable TV, internet, mobiles etc., have totally changed human life styles including for the poor. Biomedical products and pharmaceutical products have increased human life span even in poor countries like India. Agriculture has also changed over the past few decades, world over, with GM crops and new forms of irrigations.

Still problems of poverty continue. Scientists/Engineers/ Medical professionals tend to blame the policy makers (politicians, economists and administrators) for not realising the full potential of the scientific possibilities, which have been proven so well in many places. For example: Why not enable drip irrigation all over India to help all agricultural fields in India? Why not remove water scarcity through waste water reuse and recycling? Why not use GM crops? amongst many other options.

Some of the economic debates on poverty appear far removed from reality or practical actions. It is not a simple arithmetic of balancing government expenditure: that is the taxes collected and the expenditure towards" poverty removing schemes".

It is not like running a machine by giving it fuel continuously. Even it will show wear-and-tear soon after a few years.

Economics and society are more complex; complex, even more than biology, from a human point of view. Organisms in biology draw energy from its surrounding (Nature), self-organise and continue to grow. They reproduce; over a long period experiment, mutate and evolve with new features. Also there are catastrophes of nature; non availability of water (water stress) to non-availability food. Predator in biological systems cannot continue indefinitely to prey, as the population of the prey may disappear, even after "ingenious adaptation" by the predators. But under normal circumstances, there are other options like migration to new areas.

In the overall, biological, eco systems adjust and evolve, "complex adaptive systems" as they are called. In fact, human societies also have evolved in these ways, as a part of the total "biological-system-of-the earth" called biosphere.

Look at the human invention of organised agriculture. The process of agriculture itself invented and combined with many processes of accelerating "the natural evolution". Many new species of plants and animals were "cross-bred" and domesticated by the human beings. Many of these have eliminated the earlier "wild" varieties. With so many new human made, varieties, a major invention of Trade as a Societal System took place: various forms of barter deals to money transactions. The rules (or governing systems) for such systems were required. That led to organised societies with rulers, ruled, and various specialisations of human work.

Thus the complex system of economics and governance was formed to grow in about 10,000 years say about 400 generations into a mind boggling system without any idea of the linkages between macro and micro (even at meso levels).

Diversity of Options

In the book by EB quoted above, he observes that the economics (human economy) started about 2.5 million years ago by trading tools. He compares "the stone-tool- making-hunter-gatherer tribe living along the Orinoco River in the remote border of Brazil and Venezuela" with "the New Yorkers the cell-phone-talking, café-latter-drinking tribe living along with Hudson river on the border of New York and New Jersey".

He describes as to how both these tribes carry the same type and number of genes. Therefore their biology and innate intelligence are essentially identical. The Yanomamo tribe makes its own products and average income is $ 90 per person years. For a New Yorker it is $ 36,000. But it is not the 400 times more income per person per year, which makes the New Yorker wealthier and richer. It is to be captured in STOCK KEEPING UNITS (SKU's). This unit is used to count the number of types of products sold by a store. If there are five types of jeans sold it is five SKU's. In the Yanonamo economy, EB estimates ALL PRODUCTS TOGETHER ABOUT SEVERAL 100's and at the most several 1000's. For a New Yorker, he (EB) estimates the economy to be order 10 raised to the power 10, SKU's, that is, many tens of billions-that is, many thousands of Cross of SKU's.

He points out that the life style of our ancestors about 15,000 years ago (when the agriculture was just invented) was typical of the Yanonamo. Now even in India though its economy is still at a very low level compared to USA, the SKU's in its major metros will be about a few tens of crores. In the other cities and smaller towns, SKU's will be much less but still running to many tens of millions and villages it may drop down to several lakhs.

The SKU's are a reasonable indicator of diversities of the economy and consumer options, and therefore the resultant value additions. Take a simple tomato. In addition several types which may be say 10, several value added products can be made and sold in market in different packets, sizes, with different tastes because of additives, etc. Then SKU's from tomatoes may rise to 1000 or more. Metros will have all of them but villages in which they grow tomatoes will have only a few tens (with a few value added products like ketchup bottles brought from towns).

Behind the story of SKU's is not only the complexity economy at a macro level of looking, but also many micro level activities: many thousands of "value additions" to the products and processes till they reach the ultimate consumers. These micro activities are productive work, giving incomes to the workers, who give the "value addition".

Now add on many thousands of agricultural products (grains, vegetable fruits, flowers, animals, birds, fishes etc.,) including the "wastes" which are produced. How many hundreds of lakhs of SKU's will get generated?

In exchange of these, many lakhs of value added products, the agricultural economy will attract many other products: from agro-machines, to drip irrigation systems to waste management systems etc. Look at another dimension: many temples in Tamil Nadu, which were great centres of economic activities as well, a millennia ago, were remaining at the subsistence levels. Now with increased wealth in the hands of many millions of people (due to their continued work at the micro-level SKU additions in agriculture, manufacturing and services) most of these temples are now buzzing with crowds. Just in front of the temples many thousands of SKU adding products are being sold: traditional flowers, to packed nuts to pooja, materials, toys to clothes amongst other miscellaneous items.

This is an example of VIRTUOUS CYCLE.

When there are major disruptions like lack of water availability for the fields; or lack of electricity for the factories or lack of transport fuels or severe disruptions to the agricultural supply chain systems due to government controls, or other physical reasons etc., the number of SKU's will fall down. Options to the consumer will go down. Lesser number of products or options will lead to price rise, inflation. Then other distortions like excessive storage by cautious consumers to clever traders, will take place. When prices rise, product choices will go down; consumers will have to pay more to get less. Their money loses value in this 'inflationary' cycle. Their incomes remain same; often times many persons who were in the SKU multiplying processes may lose jobs and therefore incomes.

Schemes like employment guarantee or food security by giving free food or highly subsidised food gains/ food items, can at best be temporary. **The key ingredient of the economic machine from the times of the TRIBAL COMMUNITIES many millennia ago was to increase diversity of products –baskets, clothes, tools, many new varieties of plants/animals/birds-medicines etc., - and further increase the SKU's through trade.** Indian and Chinese economies naturally dominated the world during the 16th, 17th century as they had the population to produce diversities of products and exchange them globally through ships and land routes.

Years after 1750 changed the situation with the arrival of automated machines, mass production and also newer tools for the military, faster transport of ships, trains, etc. It was natural that the countries which mastered the Industrial revolution ruled the world economy and also ruled many parts of the world to maintain the supply chains for making energy, materials etc., and also to increase their SKU's in the global markets.

India and China started to acquire these industrial technologies developed in the West since their Independence. Indian and Chinese people worked on the low value adding SKU's. It is through the strength of their primary sector (agriculture, mining the ores etc.,) and the hard work of many of the traditional artisans, India could pay for the import of machines. Since more number of people were drawn into the (albeit) low value added SKU's, Indian/Chinese economy started growing with more wealth creation.

During the predominantly agricultural phase till the Industrial revolution, the number of working people (population) and the number of specially-skilled artisans engaged in the SKU generation activities, determined the wealth generation capacity of the country. After machines came, the focus shifted the not merely to population but to the number of persons who are highly skilled in science, engineering, medicine, trade, transport, military equipment etc. **So the developed world has economic military and political power, far beyond sizes as a part of the world population.**

China steadfastly focussed on acquiring these skills for a large number of its population and in giving them important (albeit at low level value addition) roles in the global production chain. So much so, a good part the SKU proliferation in front of the New Yorker is with products made in China. They are mostly with the machines made in Japan, Korea, Germany, USA etc., bought by China.

India somehow missed such a focussed approach as brought out by the 25 reports of TIFAC (Technology Vision for India 2020, (1996) and also briefly brought out in the book "India 2020: Vision for the New Millennium" by Kalam and Rajan. (1998). In the phase of liberalisation and globalisation post – 1991, (which was a very good thing that happened to India as the economy could be freed from the shackles of earlier bureaucratic- socialist SKU reducing economics), Indian economists and policy makers took only a macro-view of economic growth. They only hyped about financial figures of investment and company performance. The engineering, technological and skill strengths of people and their activities were forgotten. S&T institutions and IIT's were stand alone, expenditure items paid for from tax incomes and not considered as items which integrated with the lives of people. Similarly the industries were lumped as items of macro-economic and financial engineering, forgetting the need for their internal engineering and technological build up, which alone will help them in innovation. It is innovation which leads to high diversity of high value SKU's which can win domestic and global markets.

It is not necessary that a country or society needs to have internal strengths in all sectors having "self-reliance" (an unrealistic and unachievable goal of our socialist past, which had made Indian economy nearly stagnant for four decades). Value added billions of SKU's available in New York, are not all produced in USA. But there has to be WEALTH in USA/New York to buy; otherwise shops will close; options will shrink and SKU's will plummet.

The 'wealth' referred to above is not the government budget which is generated by taxing people and their activities nor is the total wealth of rich people. That wealth of a country comes from the income adding economic activities of people, at the micro level. If they (people) are in the value chain, they will earn; they will spend; that will create many new products/services and options (choices). Even USA which is wealthy by any standard and which provides large incomes to its people, is not immune to this reality. Ignoring large scale pro-active skilling of its people. By leaving it to the neo-liberal market forces and also being complacent that the poor can be taken care of in their social security networks, unemployment insurance etc. for long periods, USA faces unemployment problems now. So USA is trying to restrict immigration in the hope that it will stimulate local employment by US companies. This is a macro step. What is more important is to skill its people/youth and induce them to take up the challenges of low end/middle end IT based activities and make them competitive visa vis Indian, Chinese, Philippines etc., counter parts. President Obama often emphasizes this aspect in his speeches.

Therefore for India, it is all the more vital that we give attention to micro level activities, not just skilling but also other economic activities starting from the farmers who have marginal lands, landless agricultural workers, to worker of many micro-small-medium-industries. Big ticket investments, FDI in retail, big infrastructure etc., have their roles, giving a macro set up. But they will fail in the country as a whole, when they are not linked to all the diverse micro-level economic activities and the diverse set of people (who are micro-level economic agents).

This is where, the role of Complexity Economics (CE) which is emerging as a serious study in the developed countries, is VERY USEFUL. As per CE, to quote from the book by EB, "wealth must be a product of evolutionary processes. Just as biological evolution summoned complex organisms and eco systems out of the primordial soup, economic evolution has taken humankind from a state of nature to the modern global economy filling the world with order, complexity and diversity along the way."CE helps to understand the processes of wealth creation much better than the existing Traditional Economics (TE) whose deficiencies are being understood after about a century of "majestic rule" by TE. The article by Mae-wan Ho quoted earlier gives full critique TE as well. Also see an article in the Economist: "What went wrong with economics" (2009).

Therefore let us not be overly obsessed with the indicators of TE as god-sent truths and ruin the economic growth of India and deprive Indian people an opportunity to earn respectable incomes in the global economic chain. If the macro-micro linkages in the complex-adaptive-system of Economics are not addressed, it is not possible to achieve the above objective of good lives for all. CE does not indicate a flat-income "welfare"society; the evolutionary processes will lead to rich and poor. The solutions to take care of the poor are not through bang-bang taxation to feed the poor. CE processes give much better insights, which will help eliminate accumulation of poverty. The solutions are not in the Right or Left extremes of TE policies. Not just a mushy-mushy middle way, but an insightful holistic way. The book by EB gives many examples and details.

The more I read the book (this is my third revision over the past few years) I am convinced that Indian traditional wisdom of holistic thinking, when used along with the analytical processes of CE, will be very useful to find practical solutions for the Indian problems.

Complex Economics for Indian Agriculture

It is with that goal in mind I have written this chapter. If the EB book itself is about 500 pages and still is only an introduction, it is too much to expect a road map for India, derived from that book. Also such macro-road-maps are the anti-thesis of the spirit and approach of CE. The book will give us mind-set-changes and indicate avenues for exploration. These in turn, would give insights for short term actions, in a better informed way.

I would request the more mathematically oriented, model oriented readers to attempt a CE model for Indian Agriculture. Marginal farmers; rain fall variations; agro-climatic endowments ; animal husbandry; poultry; fishery; forestry; related agro based industries; transport systems; water sources; water industries; human habitats and encroachments; price controls; other Government rules; Social Technology (EB book terminology) infrastructure like cooperatives NGO's labour/Kisan associations labour laws etc., in different states; fertiliser, pesticide, seed suppliers; consumer markets; labour migration, exports etc. Fortunately many of these can be modelled and simulated in the computers, as powerful computers and related software are available in India. Many competent modellers, systems thinkers and mathematicians are available in India but are working on problems which are the agenda of developed nations, as they get better money, easy funding and recognition.

Are there any forward thinking philanthropists who support such CE studies (as Santa Fe institute gave a place for study of CE and EB's book)? Indian agriculture may be looked at differently and therefore newer insights to tackle agricultural poverty may be derived.

My main emphasis is that let us not get lost in macro slogans of Right, Left, Centre etc., and instead understand the COMPLEXITY, with modern tools and also with an attempt to use Indian holistic thinking for solutions. Those who want to look at the earlier Indian (South Asian) holistic thinking in the modern forms may refer to a book by Susantha Goonatilake,'Towards a Global science: Mining Civilisational Knowledge'(1999).

Readers are also requested to enthuse others in fields of mathematics, sociology, economics, theoretical sciences, life sciences, compute scientists etc., to start a process of CE modelling of Indian agriculture, with a view to gain insights for actions.

❒

10

Agriculture in Education

Nationwide surveys indicate that given a chance, many farmers would like to migrate to a city taking up a job there. They are tired of the day-to-day uncertainties of their village life which is tied mostly to low income agriculture (fields and livestock). In addition, the rural boys vote with their feet by fleeing to nearby cities or even far away. They prefer city jobs even though they have to live in slums because they can earn more and have more options for consumption. Once or twice in a year, they go to their home villages, take with them some goodies. A good number of them especially in north India go twice to their villages to do some help in the farming in their own families or as agricultural labourers during the harvest season.

Mostly Indian agriculture depends on ageing males (many crossing in their sixties) and women who stay behind. If such trends continue there is going o be a shortage of experienced and skilled work force in the agriculture sector. Let us not imagine that most of them are illiterate and therefore unskilled. Their knowledge of detailed nuances of various phases of plant growth, insects, soil conditions, suitability of weather, or the health of animals and birds cannot be pooh poohed as traditional "low" knowledge. They are very insightful and over a period blended their family/person embodied knowledge and skills with modern agronomic practices. The capability is not only for the adopting the green revolution but also for absorption of various new agronomic gadgets and agricultural implements. They have also adopted the modern health care practices for animals and poultry whenever they are accessible and affordable.

It is also good to note that the educational opportunities for school education are increasing in villages. Girls are also going to schools. The farmer-parents are also encouraging their girls to study in schools. I have seen many persons who come from Bihar, UP, Madhya Pradesh and Rajasthan to Delhi, Gurgaon and Noida and work as drivers or in many other small sector services jobs. Most of the men are alone and leave their family back in villages. They go away to villages twice or thrice in a year for staying for a month or more to be of help to wife who takes care of the farm. The availability of mobiles has helped many of them to be in touch throughout the year 24x7. They can make an extra trip during emergencies.

When I talk to them, most of them would want their daughters to study as much as they can and get in to better jobs. "Sir, my wife and I struggle to earn more so that my daughters and sons can learn more and get into better paying jobs".

Apathy to Farming

As for the urban people, farming and agriculture occupy only a very small part of their mental space. Political persons, mostly offer lip sympathy and solve only some post crisis situations, that too with some palliatives. A full systemic look is missing. So it is with intelligentsia. Whenever I talk to school children or those in colleges, when I ask a question "How many of you would like to take agriculture as a career?" no one raises the hand, even in hinter lands of India. These children are looking at the education as a means to get away from poverty. Those who are relatively better off think of high paying jobs in IT sector, financial sector or going abroad.

Food availability and supply of low cost food items are taken as entitlements by urban people – from the lower middle class upwards. When I talk to intelligentsia about moving out the surplus work force in agriculture (about 60 percent of the Indian workforce being in a sector which produces only about 15 per cent of GDP), their immediate response is "How can they be moved out? They need to take care of agriculture". The very same people will also want zero-inflation in agricultural prices: be it grains, vegetables fruits, eggs, chicken fish or mutton.

Urban people, having the ability to gather in large numbers even in some part of a city can create "protests"; politicians are more attentive of their demands. Media can project such protests easily. I remember how during the 1960's, the trade unions used to paralyze cities or states, through a call of strike for increased rations or against rise in prices thus demanding more salaries for them.

What about farmer's increased income? Do they have to work on zero increments life- long (even from generation to generation?)

Few think of these issues. We have been raising these issues in this series, asking for "agricultural prosperity".

Dr. N. Mahalingam has been raising this issue in the Kisan World for the past four decades: making agriculture, that is being a farmer, as respectable (and well earning) profession as being a doctor, engineer, scientist, academician, teacher, banker administrator etc.,

We have been arguing and pleading for reforms and liberlisation in agricultural economy; these are vital. So also is the need to create in the transition a large number of rural based industries (MSME's) to absorb the off loaded work force from the current low productivity agriculture. Allowing bigger players in agriculture (farms, animal husbandry and poultry) through lease-and-contract processes is also important. Bigger supply chains that will be built in this process of transition will absorb many persons out of low income, agriculture, **leaving it as a place where bigger players can generate more efficiency and productivity of the land and thus more wealth. Then farmers including those who lease the lands to the bigger organized players will become wealthier (earn more income).**

Even if such processes with the necessary legislation are put in ground in 2016, it will take about a decade to fully effect the transition.

Due to the continued neglect of and apathy towards agriculture and farmers especially over the past four decades as the emphasis and focus shifted to manufacturing(to the neglect of agriculture) and then to IT and services sector (to the neglect of both manufacturing and agriculture), the better talents have moved away from agriculture. Even children of rich farmers have moved away from agriculture. They got educated, went abroad. Some returned to become a part of private sector (mostly non agriculture) – post 1991.

Shift of Talents

When we start retrospection on the demographic shift of the Indian talent force, one would notice that about a millennia ago, many of Indian traditional arts, culture, and religious activities attracted the best of the talent. In addition those positions where one could command a lot of land and trade channels, were sought after. With the coming of the British, the talent migration was towards education in English to occupy powerful and better income jobs in the British governance systems. Others who did well were those who could control farming through the feudal zamindari systems and the foreign trade.

Post-independence, the talent shift was towards Government jobs, Government industries, government created higher educational system, allopathic medical profession etc. Of course politics attracted many persons. A little later, the industrialists in the private sector also started getting respect due to the policies of mixed economy, though constrained by the license-permit-quota-raj. Land-to-the-tiller and other land ceiling Acts gave a temporary relief to the farmer. The green revolution placed the farmer closer to the centrespace especially after the country was battered for over a decade with severe food shortages and the resultant urban unrest.

Once food crisis went away, the focus shifted. No doubt, it is natural for any modernizing economy to grow more wealth in manufacturing, infrastructure, mining, services etc. Therefore the share of agriculture sector went down rapidly. More people trapped in a sector whose share in the national kitty is going down towards 15 per cent.

World wide experience (especially all of the developed world) faced this phase with a rapid growth of manufacturing sector going upto 35 per cent of GDP share. In that also a good mix of big, medium and small enterprises were there to absorb the excess labour released from agriculture. Over a period they acquired several new skills and went up the ladder of talents in many technical support areas. Then came the dominance of the services sector.

In India the agriculture was going down in share (note not in absolute physical terms). But manufacturing did not go up rapidly. With the globalization and external competition, our MSME's took a beating. Even bigger manufacturing enterprises depended upon many imports which also cut down local jobs and a natural supply chain to MSME's. A number of agricultural labourers (after part time work during peak agriculture seasons) went away to towns in low level services sectors: domestic works, petty trades as casual workers. Many of their agricultural knowledge and skills are of no use in these jobs. The Indian services sector is a mixed bag of expenditure on Government employees (a large chunk), a huge low level unorganized services (many people involved but wealth wise low), education (middle income levels) as well as high income levels of IT and finance sectors (which employ only a very small segment of work forces). The talents move to the latter, as it needs to pay only a few people with a huge wealth created. Next are government jobs but they are not expanding! The demand for smaller states is also due to the fact that it will create more Government jobs which are well paying in the Indian context and secure.

In this milieu, how can talents be retained in and attracted to the agriculture sector? Even when one can imagine a highly corporatised or cooperative run (on commercial lines as Amul), agricultural businesses along with many rich and medium scale farmers with the own operations (in this scenario marginal small farmers would have leased their lands to the cooperatives or corporates or rich farmers). There will be a requirement of highly talented agricultural professionals and agricultural workers in the field. Agriculture cannot be all mechanized, all computerized systems, even when Precision Agriculture is adopted in some large farms (we have discussed Precision Agriculture earlier). In addition to the regular operational talents for agriculture planning, operations, and supply of products; waste management; waste water recycling and use; supply chain management for inputs and outputs, there will be a need for continual research, and innovations.

In India, we lost out the momentum of the Green revolution because we were complacent. Nature is never complacent. If we increase our productivity, pests and insects do too. They also try to beat our pesticides, creating new breeds of resistant pests. It is a continual battle. Also soils need continuous care. We have had glimpses of soil biology in an earlier article. Then there are other non-biology related items: people's desires and demands change. Fashions and fad change. Micro economics change. These are called the 'market forces'. Thus agriculture as business will require continual monitoring of these: global trends (we need to export as we are richly endowed country for agriculture), domestic trends, new opportunities etc., for example Genetically Modified Maize (GM Maize) based ethanol may become an essential component of Energy Mix required for India at some part of time. Or Africa or Middle East may need certain grains or vegetables or meat products. Ability to sense market forces and succeed in business terms requires many forms of integration of domain knowledge of agriculture, capacity to deliver etc., All these will require highly talented persons (from researcher to technician workers to modeler).

If after policies for a liberalised agriculture are in place, we may be in trouble. The Defence sector is a classic example. For about two decades, there have been demands to have policies and procedures for offset production and private sector participation in the huge Defence Procurement. I have had a part in working with Confederation of Indian Industries (CII) and Ministry of Defence (MoD) to evolve policies and procedures. It was a detailed exercise involving all stake holders. There were seven volumes of reports. It was around late 1998-99. Central Government was pro-active (that is why the reports were generated) and was willing. But the unfortunate Indian time-rate-of-interaction (my terminology coined around 2000 to describe the slowness of our systems), ensured that nothing concrete took place. Engineering capabilities are not built up on a gazette notification or a Government Office Order or a call for an auction.

After more than a decade, India became the Number One in Defence Import. For the offset policy (that is the company from which India imports has to do 30 per cent outsourcing from (local) Indian companies) **does not work on ground. Most**

Indian MSME's do not have the capacity and capability to supply such offset-outsourced-items. They did not have the opportunity to build up their skills and talents. One industry person had told me earlier "Sir: Kurudanaikondu raja muzhi, pannuendrusonnal, yeppadimudiyum?" "Sir, how can you ask a blind person to pose with a great Royal Look?" The current good push for Make in India is facing this major hurdle. That is why Government is encouraging many joint ventures with foreign companies.

A senior industry person quoting the above to me(2000) in the context of a breakthrough and innovative R&D proposal to be executed in an Indian industry for which part funds were available for Government of India. We had worked out many technical details required for the proposal. But to execute it, one would require a good team of persons experienced in such R&D works earlier, at least at a small scale. Instead all our engineering and technical personnel – be it in private or public sector industries – were prohibited or positively discouraged by the same top management (which talked of technology/R&D in public platforms in late 1990's) not to tinker with the processes or procedures given by the Principals (their "god" supplier of technology equipment and consultancy). The senior industry person, who himself was an engineer, told me that if any engineer insisted on doing some small changes as that would increase productivity or improve ease of operations thus reducing rejects in manufacturing, the standard grudging reply was "Then send your whole plan to the Principals and get their written approvals". Some did so. The Principals gained through such a "free knowledge", in turn they charged the Indian company for additional consultancy fees.

This was how the Indian industry grew up right from 1947 (barring a very few exceptions about which also I am aware of). So the industry person explained to me "Sir, we have thus made generations of engineers and technicians "blind persons", a forced blindness! How can they suddenly give a Royal Look, just because the top losses now talk of R&D in industry and government is ready to give part funds?"

That is the reality about every knowledge and skill. It is not enough to acquire but should be exercised regularly in the actual working situation; then try small changes/innovations; absorb new knowledge and skills.

This episode came to me in the process of my retrospection. For over four decades, the Indian youth, (urban, rural, rich, middle class, poor or whatever classification) have been given a feeling that Agriculture work is in the bottom of the pyramid!". 'Your own future is elsewhere' is the subliminal message they received. Post 1991, things became worse; even manufacturing was looked down upon. Given a chance, all Indian youth would like to sit in front of a computer console in an air conditioned room and keep working. They will be doubly happy if they can "wire in" their ears with a mobile set, for music or talk.

When agriculture reforms do take place (it is already overdue for many years and I expect it very soon just as 1991-liberalisation took place) – we may have to face the same situation in agriculture sector as narrated above for industry: existing persons trained under multiples of constraints, the experienced ones getting old and talented young going away from agriculture.

Catch them Young

Therefore we need a massive campaign to attract talented youth towards taking a profession in agriculture: as farm operators, managers of supply chain, manager of agri-operations, animal husbandry farms, poultry and fisheries; also variety of specialized operations in vegetables fruits, special cereals, cash crops, export type items, mass consumption items, pulses, farm machinery operations including loaning, servicing etc., Of course there are higher end operations like tissue culture on commercial scale and supply to farms; preparing hybrid seeds; GM seeds; customized supplies of fertilizers and minerals to suit soil/crops combinations; organic farming for special markets; flowers for exports and domestic use, etc., (List is just a glimpse. One has to get into the super specialisations in order to derive maximum profits and market share).

Such an attraction cannot be done merely by the admission to agricultural universities. They are not the top of choice for our youth.

In order to enthuse them we have to catch them right from elementary schools (1 to 8th class) and going to secondary schools. Without loading children with another set of "rote-lessons", interesting Indian agriculture related lessons, videos, picture books, etc., should be given as apart of compulsory learning for children.

I am normally against anything made compulsory. But the apathy towards agriculture as a profession is so high that there is an urgent need to reach out to children very soon. I am also not unaware that if the lessons are not done creatively and interestingly, it may turn away children from agriculture furthermore.

It is essential not to archetype agriculture education to children, as is mostly being done for the compulsory environment lessons. It is almost reduced to planting trees; or talk of organic; or solar equipment or against plastics. Mostly they give the child an orientation against any industry, without giving the really balanced picture.

For agriculture, the message to Indian children in schools should be: it is important for everybody; many exciting new things are possible and are being done in India; how the lifestyles of those who have taken up full time modern agriculture is equally good.

There are many good examples all over India about some young people taking to agriculture and succeeding (despite the ecosystems of governance not being supportive). In between the great potentials that exist can be shown through Israeli examples. Children will see how many technologies are brought together; they should also see how profitable the business is. **And that they can aspire to be wealthy through this profession as well.**

You Tube type short videos can be made in all Indian languages promoting this knowledge. I hope the TV media will also pick up such success stories and enthuse the children. Such a sensitization should not be limited only to telling about the managers and investors. It should cover a spectrum of other middle level processional; para-agricultural staff who are in the modern farms in India and say in Israel or Taiwan etc.,

In a nutshell the message: "Children, you have enjoyable jobs, incomes and lifestyles by working in modernized agriculture. Don't turn away seeing only the reports of poverty, dirt, drought, misery etc. There is another unreported side you should be aware of. The India of the future will rapidly move towards such agriculture as our country is specially blessed for agriculture. India can, not only supply quality agri-products for domestic use but also emerge as the number one exporter of many quality agri-products.

Even before government machinery moves, some progressive schools start attempting to have such educational material introduced to the children and also take them for picnic or field visits to such places. I appeal to readers to enthuse schools, and creative video makers to take up this challenge of teaching in elementary and secondary schools and also in colleges B.A., B. Com, B. Sc, BB. A etc., we need many lakhs of talented youth to turn to agriculture as a profession.

❑

11

Electricity from Renewable Sources

I had chosen to write on the above topic even when I was writing for the April 2014 issue "Chaotic Churning: What is ahead?" In a two decadal horizon, **electricity from the renewable energy will become important.** But that will be so only when we solve the current day problems of poor and erratic electricity supply to people, industry and agriculture.

If one looks at the website Open Governance India (2014) under the section India Regional Electricity Statistics, India is still far behind in terms of reaching electricity to all its citizens. Access to electricity in terms of **percentage** of people let us see some highlights:

Afghanistan	30
Algeria	99.4
Argentina	97.2
Bangladesh	59.6
Bahrain	99.4
Brazil	99.3
China	99.8
India	75.3
Indonesia	73
Malaysia	99.5
Pakistan	68.6
Saudi Arabia	99
Sri Lanka	85.6

The first important step to remove poverty to a person or a family in the modern world is to provide a good electricity supply for them. It is needed to pump water. It is needed to provide access to modern knowledge tools starting from TV to computers. It is essential to have a modern competitive industry for a micro-or small industry. It is needed to increase agricultural productivity or artisanal productivity. I can list many more.

The above given 75.3 percent coverage of the people in India, hides an important information!

If we look at the rich source of information available in Wikipedia under the title Electricity sector in India, about 800 million Indians still use traditional fuels (fire wood, agri waste and biomass cakes) for cooking and general heating needs. Particulate and other pollution from these are the reason for death of about 300,000 to 400,000 people in India annually.

Note this important statistics: Some 400 million Indians lose electricity access during the black out (amongst these 75.3% who have access!!). While 80% of the Indian villages have at least one electricity line, just 52.5% of rural households have access to electricity! In urban areas it is better; it 95.3%. No wonder rural people want to come to urban areas, even to live in a terrible slum.

Also even where electricity access is good, the outages are very bad in most States. That is one reason for the economic slow-down and loss of jobs.

Amidst this disturbing (if not dismal) state of affairs for one crucial element, ELECTRICITY, for living a modern life, how do I think of decades ahead? That was gnawing me even when I was reading up lots of material on Electricity from Renewable Resources to write the July 2014 article.

One intuitive conclusion (and perhaps hope), I had arrived at the April 2014 article (now in the chapter 12) was about history being made in a few weeks of chaotic churning of April-

May 2014 in India alluding the Elections 2014. Now India can hope for a decade of stability, which is a good enough period to correct most of the 'sins of the past' and the gaps due to missed opportunities. We had written about these in the earlier chapters.

The new hope and a massive mandate for stability, gives me confidence to share with the readers an important aspect of the long term future of the world and India, that is to be able to source electricity from renewable resources. **Such a transition is possible only when in the near term we are able to give access to quality electricity to all Indians say within 5 years and thus start increasing their income levels** (India's GDP will grow fast with such an access and other focused policies for agriculture, manufacturing, etc.,). Then the population will be prosperous enough to switch over from the 'cheap' electricity with the fossil fuel technologies.

The Present Electricity Sources

Even the well-meaning and well informed activists for renewable sources get carried away by its prospects so much so that they all lobby against coal. That is the reason India has a serious electric power crisis all over India and also slow- down of various sectors.

Let us look at some statistics from the India. Regional Electricity Statistics quoted before (Numbers are percentage of total)

Source of Electricity	Installed Capacity	Actual
Steam	53.14	67.39
Diesel and Wind	8.62	3.95
Gas	10.27	9.65
Nuclear	2.88	2.35
Hydro	25.10	16.66

Steam is the bulk and most steady supply; it is mostly of coal and other petroleum fuels. But India is not producing enough fuel/coal to feed its power plants. Some plants do not have reserve coal supplies to last a day of operations. The monopoly government owned Coal India had been resisting modernization. I am aware it since my working with TIFAC (Technology Information Forecasting and Assessment Council) since 1988 which was advocating clean coal technologies. In addition, the current environmental awareness level of people leads to resistance to miners' intrusion into forest lands (where many of Indian coal reserves are found). Also compared to the average emission levels of the plants in European Union (EU) 27 countries India's thermal plant emissions emit 50% to 120% more of CO2 per kwh (kilo watt hour) of electricity produced. Thus, in the short term there is lot that needs to be done in the coal and other thermal plant fuel **to ensure that the bulk of electricity to India is delivered.** Any slow down or dithering on that front would be detrimental to India's (and Indians')economic and social progress. (Fortunately after the unprecedented order by the Supreme Court, the government has moved fast to auction and allot mines. More needs to be done to liberalise coal sector).

While we can pursue the nuclear route, it has lost out due to its slow progress over the past several decades and it is unlikely that its share will be any significant. In fact renewable energy electricity may become more than the nuclear share.

Hydro power also seems to have saturated due to a variety of slow actions in the past. Now the main focus may be on improving its down time and make it a steady power. If major river linking projects come up in future (unlikely as can be seen now due to technical, environmental and social complexities) hydro power share may go up.

Other Elements for Electricity

It is not enough to look at installed capacities of the sources of electricity, and generation of electricity alone. **The final 'proof of the pudding' is how the generated electricity reaches the ultimate consumers.**

Let us look at the data as given in the India Regional Statistics (2008) in giga watt hour (GWh)

- Generated Power 677,094 GWh
- Consumption 501,977 GWh
- Loss in Transmission 187,513 GWh

We are not complicating this information by further (important considerations) of available base load and available peak load. The data is available in the Wikipedia source quoted earlier. As given in that report, **"The gap between availability and demand is the shortage which India is suffering". Of course "this shortage as in these statistics ignores the effects of the waiting list of users in rural, urban, and industrial customers". "It also ignores the demand gap from India's unreliable electricity supply."** This unreliability and gaps in transmission are not uniform all over India. Some States in India have overcome these difficulties through proper policies, governance, correct implementations and continual monitoring. But many other States suffer.

The fact that India and Indians are electricity starved has not become a political issue yet. It is all around the decades-old issues of 'food-starvation', and 'doles'. Hopefully the new government with a massive mandate for a new mind set will solve these problems, which are easy to solve, given the political and administrative will.

As given in Wikipedia "Average transmission, distribution and consumer level losses exceeding 30% which include auxiliary power consumption of thermal power stations, fictitious electricity generation by wind generation and independent power producers (IPP's) etc."

So India's power 'losses' are also due to fictitious generation statistics, a clear indication of corruption in governance. Therefore the solutions are not merely scientific and technical.

But India needs to bring down the technical power losses through better technology indication. The 'fictitious' accounting can also be taken care of through tamper proof metering and severe punishment to those who 'cheat' and are cheating.

Renewal Source for Electricity

As we have discussed before, India from now on (immediate) to coming several years, has to concentrate on conventional thermal energy sources, coal being the dominate, gas being the next best to help meeting the peak load demands. However, coal use need not be in the same old environment unfriendly and non-clean mode. Coal mining also needs lots of safety systems.

Though India has been and is being rocked by coal-scams, one cannot depend on the monopoly of the inefficient and monolithic public sector Coal India. Private and foreign investors are required to bring new, clean, safe and competitive technologies.

Let us remember the current statistics (Wikipedia): "the coal fired plants account for 59% India's installed electricity capacity compared to South Africa's 92% and China's 77% and Australia's 76%." In the developed nations where coal is lesser, it is the nuclear power electricity that takes a giant share (example France, Japan even USA), we noted that India is unlikely to pick up nuclear. Hence coal based electricity is a must and let us all support it.

Remember the earlier statistics about Hydro, Gas and Renewable. Therefore, whatever we discuss, hope for the renewables should be seen in this context.

At this juncture, we would like to draw the attention of our readers to an excellent report by the National Academy of Science (NAS) of USA titled "Electricity from Renewable Resources:

Status, Prospects and Impediments". It can be down loaded from website for nominal charge or free. Naturally the report is in the context of USA and fully relevant to it. But others can learn from it. In this report, renewable energy excludes hydropower, as it is conventional and known. It concentrates on solar, wind, biomass, geothermal etc.

The report addresses in detail the resource base, renewable technologies, economics, environmental impacts, deployment, scale of deployment and the integration of renewable electricity. It also discusses the future prospects of renewable energies and the critical unknowns. We do not discuss all of these in the Indian context; I have not been able to see such a holistic and yet in depth document in the Indian context. It will be good if some institutions and researchers in India first attempt such a report.

The major findings of the report:

"For the time period from the present to 2020, there are no current technological constraints for wind, solar photo voltaic and concentrating solar power, conventional geothermal, and bio power technologies to accelerate deployment. The primary current barriers are the cost-competitiveness of the existing technologies relative to most other sources of electricity (with no costs assigned to carbon emissions or other currently unpriced externalities), the lack of sufficient transmission capacity to move electricity generated from renewable resources to distant demand centres, and the lack of sustained policies. Expanded research and development (R&D) is needed to realize continued improvements and further cost reductions for these technologies.

"It is reasonable to envision that, collectively, non-hydro power renewable electricity could begin to provide a material contribution (i.e., reaching a level of 10 percent or more, with trends toward continued growth) to the nation's electricity generation in the period up to 2020 with such accelerated deployment.

"In the period from 2020 to 2035, it is reasonable to envision that continued and even further accelerated deployment could potentially result in non-hydroelectric renewables providing, collectively, 20 percent or more of domestic electricity generation by 2035."

"In the third time frame, beyond 2035, continued development of renewable electricity technologies could potentially provide lower costs and result in further increases in the percentage of renewable electricity generated from renewable resources. However, achieving a predominant (i.e., >50 percent) level of renewable electricity penetration will require new scientific advances (e.g., in solar photo voltaics, other renewable electricity technologies, and storage technologies) and dramatic changes in how we generate, transmit, and use electricity".

The report lists a number of areas in which scientific advances need to be focussed.

We have quoted the above **in extenso** because, USA which is a scientific and technological giant in renewable electricity technology (with a large share of patents in the world) is only able to take such optimistic targets. Therefore in India where we are yet laggards and followers, we need to be cautious in goal-setting for renewable electricity use in India. One cannot gloss over cost-inefficiencies by subsidies.

Again a word of caution: one should not get carried away by the statistics of installed capacities. One should look at the generated power and delivery. Typically wind power electricity will be 30% or lower of the installed capacity. (See the overview statistics given in the earlier part of this article). This is because wind power is not available 24x7. So solar power can be a good alternate option as most parts of India have a vast availability of sunlight.

Renewable Electricity in India

From the Wikipedia data, we note that as on 31st January

2014, India had an installed capacity of about 31.15 GW (giga watts) of renewable technologies based electricity. Most of them are in grid connected power.

Technology	Installed capacity in MW@ 31 January 2014
Wind	20,298.83
Small Hydel Power	3,774.15
Baggage co-generation	2,512.88
Solar	2,208-36
Biomass Power and Gasification	1,285.60
Waste to Power	99.08
Total	30,177.90

Off Grid Power are about 973.13 MW, with bagasse cogeneration 517.34 MW/SPV system (above 1 KW) 159.77 MW Biomass gasification (Industrial) 146.40MW/Waste to Power 119.63 MW Biomass gasified/Rural) 17.63 MW etc.,

Solar Power

Wikipedia quoted before gives a comprehensive summary of many new recent initiatives by government of India as also some vital statistics. We give only a few glimpses.

India's solar energy potential is one of the highest in the world: energy of about 5000 trillion KWh is per year is incident on the India's land mass. Most parts of India receive about 4.7 KWh per square metre per day.

Government's Solar Mission is intended to capture some of this immense potential: about 20 GW grid based power, 2GW off-grid solar power. It plans to cover 20 million square metres with solar energy collector by 2020. Also India plans utility scale power parks. Some state Governments like Gujarat and Rajasthan have taken major lead in these. Prime Minister accords high priority.

However, solar cell production in India lags behind. Lots of solar panels are imported from abroad, major share being from China due to very low costs. Also for the long term, It is essential to have most advanced Solar Photo Voltaic (SPV) technologies if renewable energy based electricity has to become sustainable. We briefly address this aspect in a separate section of this article.

Wind Power

India has the fifth largest installed wind power capacity in the world. In 2010, wind electric power was 6% of the India's installed capacity and 1.6% of the country's power output. Tamil Nadu accounts for 30% installed capacity, followed by Maharashtra, Gujarat, Karnataka and Rajasthan. The State of Gujarat is estimated to have the maximum gross wind power potential of about 10.6 GW (Ref Wikipedia).

Biomass Power

India is rich in biomass and has a potential of 1688/MW (agro residues and plantations), 5000 MW (biomass cogeneration) and 2700 MW (energy recovery from waste). Biomass power generation as an industry attracts about INR 6 billion (Rs. 600 crores) every year, generating more than 5000 million units of electricity and yearly employment of more than 10 million-man days in rural areas.

GEOTHERMAL ENERGY AND TIDAL ENERGY has scope in a few areas.

Main sources are SPV, Wind and Biomass (and limited small hydels).

India has to concentrate on these. Biomass and urban wastes are not yet fully tapped. If these are attempted in off-grid mode, there is a scope to cover the 300 million Indians who are yet totally unreached. If one can also attempt hybrid systems with some coal wastes etc., (wherever there is a paucity of availability in biomass supply). In fact small SPV's along with biomass generators (with possibility of mixing coal) etc., should all be considered for feeding villages, locations in cities where electricity supply is non-existent or weak.

In the Indian context where access to electricity is not there for many millions of our people (about 30%) and where many more suffer from poor supplies even for their agricultural and industrial consumption, the share of renewable source electric power should not be counted as statistics of share. We need to aggressively pursue many decentralised stand alone (off-grid) renewable electricity supplies even as a transition for next couple of decades.

In the meanwhile we should not be satisfied merely with the low efficiency solar SP and SPV systems being deployed in India.

Advanced Solar Power

We therefore will have a quick glance at the reports of the Fraunhofer Institute for Soar energy systems ISE. Fraunhofer ISE (FrISE) makes Annual reports. I have had the pleasure of reading them right from 2011. The latest one is 2013-14. The FrISE approaches the use of solar energy comprehensively and holistically.

They cover energy efficient buildings; silicon photo voltaics; III-V and concentrator photo voltaics; dye, organic and novel solar cells; photovoltaic modules and power plants; solar thermal technology; systems integration and grid – electricity, heat, gas; energy efficient power electronics; zero emission mobility; storage technologies; and energy systems analysis. Each of these titles carry detailed description of achieved successes and activities aimed at addressing the gap areas in order to make SP systems commercially competitive with other traditional sources of energy. These are also covered in the USA-NAS report.

The FrISE report gives many technical and scientific details, to help those who want to do more (including the use of FrISE knowledge). I skip most of it as it is available in their website. I give a few quotes which will make the readers understand how focussed they are **in defining a R&D problem** (unlike what happens in India now).

- "Around 85% of manufactured solar cells are based on crystalline silicon. The efficiency, price-to-performance ratio, long-term stability and potential for further cost reduction indicate that this peak performer in terrestrial photovoltaics will continue to dominate the market in the future. Our activities mirror the complete value chain for crystalline silicon photovoltaics"...... ...."for highly efficient solar cells and producing wafers with little kerf loss. We are developing specific equipment............"

Many details are covered over about 20 pages. While discussing wafer bonded III-V multi junction solar cells (which need to be used more in future), see the quote:

"The higher the efficiency with which we generate this electricity from renewable energy sources, the smaller is the amount of land area and the material that is required............. These special solar cells are used in highly concentrated sunlight................"

Look at the focus! As solar-cells-derived electricity is used (we have low to medium efficiency SPV panels in India now) not only the silicon is wasted in lower efficiency cells **but also the demand for land area will go up.**

If renewable-energy-based-electricity has to be used in a major way in future going beyond 20% of the total power generated, the land used will also become a limiting factor, as it is now for hydroelectric power.

Our researches have to be focussed in such a manner in the Indian context. To live with Nature in the modern world, we must use less and less material; less and less land space to get more and more electricity. Let the advocates of renewable energy remember this as well.

Conclusion

There is no escape in the long term for Indians and India from using renewable sources based electricity. Whatever India is doing now for renewable-based-electricity is good but all of them will require rapid technological upgradation so that the renewable-energy-based electricity is competitive. In the meanwhile, coal/gas based power plants are required to take care of most of the electricity needs of a growing India (which is starving for good electricity access). Let us use in those plants as well with modern clean technologies to make coal/gas/fuel pollution the minimum. Also the mining areas need to be "cleaned up" and made safe.

All of these are possible with existing well tried out technological tools available in the world.

In order to look beyond 20 years or more let us work (right now) with focus on future highly efficient renewable electricity on the lines addressed in USA-NAS report and the Fraunhofer ISE annual reports.

❐

12
Bioprocess and Bioproducts–Emerging Trends

Current media discussions (in June 2014) are around inflation in food articles, effects of bad monsoon such El Nino among other natural calamities. Plenty of worst case scenarios are built up. Indian agriculture is always depicted as one in crisis: poor monsoon; farmer suicides or inflation especially in onion prices or hoarding etc., The upper middle class which "controls" media (print and electronic) has very little connect with the Indian agriculture especially the farmers (rich, medium or marginals) or farm workers. Farmers are seen as a part of greater macro-statistics or a person shown in TV with parched faces, and often a figure that will invoke sympathy.

For politicians, they are seen as a vote bank; but they have proved themselves differently during the recent parliamentary elections. They have shown that they have intuitively understood the realities of modern globalizing economy and its realities! They seem to have understood that bulk of them, about two-thirds of Indian population, cannot be condemned to live in a marginal, subsistence economy with slogans of "Jai Kisan" of the 1960's. Nor the modern day dole of "free electricity"which seldom flows into their pump-motors. Nor the ever increasing minimum support prices (MSP's). Farmers and farm workers being close to nature and therefore the realities of soil, water, pests, etc., know the self-limitation of life. They don't have to be given lectures on "Complex Adaptive Systems", which Nature is, therefore the economics are, and hence, markets are!

As we had addressed the futuristic scenario for electricity from renewable energy, even when we have to be firefighting with coal supplies, erratic electric power supply etc. in the earlier chapter. In this chapter we wish to point out issues relating to the medium term and long term future for agriculture and plant related products for the economy.

What all we have been pointing about agriculture sector to liberate it from the past and currently continuing controls, are vital. But we should not get locked with the past and the present firefighting.

Monsoon Proofing

In addition to the above mentioned items, immediate short term action (starting right from now) is to monsoon-proof Indian agriculture. I recall a reference on file from the then Prime Minister Indira Gandhi to Prof. S. Dhawan then Secretary, Department of Space/Chairman, ISRO, in 1976, about his considered views on variability of Indian monsoon. Since monsoon was erratic during that period for several years (no satellite picture, which ISRO satellites and foreign satellites provide now, were available then!) there were lots of loose hypothesis that Indian monsoon pattern has changed. Ministry of Agriculture also felt so.

Since the file reference came, I as Scientific Secretary ISRO, had to advise Secretary DOS. I read an excellent report by (now late) Prof P.R. Pisharoty about 100 years of Indian monsoon and also a cyclostyled copy of excellent seminar proceedings on variability of Indian monsoon conducted by the Physical Research Laboratory (PRL), Ahmedabad. The net findings of both these reports were that it was very difficult to decipher a pattern of Indian monsoon in detail. The long term average is a very gross measure taken over all of India (a huge geography) and also over a long period of end May to end September. Rainfall before May end and after September and these are not factored in as monsoon. It is not their fault. Scientific measurements put some limits and follow those limits for the sake of year-to-year comparisons. So we had given all the details and advised the Prime Minister that it was difficult to conclude about definite changes in the Indian monsoon.

Such "hypothesis" of variability went on in the later years even after availability of Indian satellite INSAT images and several other satellite inputs. Various regression models were attempted. Super computers were purchased to accurately provide agro met services. Still accurate predictions even at village levels are eluding. We still chase the gross parameters like El Nino. (Once, snow fall in Europe was considered important!)

I am elaborating these only to point out how difficult is the science of monsoon. Even if some day in the long term future, we can have some accurate prediction at a village level, it still will be difficult to control rains! We have to live with what Nature gives. If we can get an advance prediction of excess rainfall (or down pour or cloud burst) some precautionary measures can be taken.

As regards agriculture for individual farmers even with very large farms or a group of farmers in a particular village, the predictions of long term national averages are of little use. Agriculture requires accurate advance predictions over the full period of plant phenology. Given the science of predictions, as it is today, it is a tall order.

Therefore the only course available is to protect the plants from "Water Stress" which occurs due to non-availability of moisture at the right stages of phenology. About two-thirds of Indian agricultural lands suffer from this uncertainly as they are non – irrigated and 'rain fed'. That can be solved through micro-irrigation systems and drip irrigation systems. These are needed for most of the arable fields in India. These have to be provided with a large part of public (government.) funds, to be operated locally with their expenditure: Build- and-transfer for operations. We won't repeat these items again.

But we need to look beyond this survival mode to reasonable incomes. If such systems are put in place in short term (3 to 4 years) Indian agricultural prosperity will begin to bud.

But more can be done. As well researched by the TIFAC (Technology Information Forecasting and assessment Council) in its report on Technology Vision for India 2020 in the Agriculture sector, and described in brief in the book "India 2020: Vision for a New Millennium" (1998) book by A.P.J Kalam and Y.S Rajan (1998), India does not even need the currently used 170 million hectares to produce grains, vegetables etc., to meet the growing domestic demands. India can produce about 400 million tonnes of grains etc. in about 100 million ha of arable land. The 400 million tonnes of grain will not only meet the demands of direct consumption but also the indirect consumption through poultry, animals, fish etc. In addition about 50 million tonnes will be available for exports. Of course this will require induction of many new technologies, GM crops, and newer agronomic practices.

The released 70 million ha need not be given away to non-agricultural use. Modern society and global economy demand much more from land, than being the basic producer of food. The fossil fuels (which are indeed from the trees of the past several million years) are depleting; nuclear fuels are limited. Hydro power sources which are real renewable energies are also limited. Solar power is not that easy to tap. Wind power is intermittent. Plants are the great tappers of solar energy, very efficiently. Therefore any additional available land will be in demand for renewable energy sources from plants – bio products through bio processes.

These are the medium term demands from agricultural lands. Agriculture will transcend the narrow view of FOOD SAFETY. It will expand to areas where it will give ENERGY for transportation, electric power, etc., and also give many industrial products, let alone many medicinal plants and aromatic plants.

Bioprocess and Bioproducts

Medium term means about 7 years. To achieve results in that period, action should start now. Many of the developed world countries and even the developing ones had started taking concrete actions several years ago. They have a lead. In India, we have specialized on talking more and doing less! Therefore we would like to point out a few areas in which our specialists and business persons can start actions, right now.

There is a good report brought out by TIFAC (2009) with a title "Bioprocess and Bioproducts" – Emerging Trends. The report is edited by Soumitra Biswas, Adviser TIFAC/DST, P. R. Basak and Nirmala Kaushik, both. Principal Scientific Officers, TIFAC. Their e-mail: biotrifac@gmail.com, TIFAC website is www.tifac.org.in

The report comprises five major chapters: Bio-ethanol, Bio-chemicals, Bio-oil, Bio-hydrogen and Industrial enzymes for Bio-ethanol from Ligno-celluloses, written by a number of eminent experts with vast experience. The chapters have been reviewed by independent experts. Technology and research minded readers as well as entrepreneurs/business persons who would like to enter into new businesses of the near future are recommended a detailed reading of this report. It not only contains information about the scientific processes and technologies but also techno-economic considerations as well. The report also clearly lists the areas in which there is still a knowledge gap and gives a possible direction to gain further knowledge for commercial success..

We quote some parts from the Executive Summary of the report.

"Effective utilization of biomass for its conversion into energy and value-added products assumes priority for developing feasible alternatives to convert biomass into liquid & gaseous fuels, chemicals, biomaterials etc. need to be considered with adequate emphasis in Indian context. There exists a wide spectrum of technology development possibilities in this area. The process parameters, yields etc. with reference to Indian biomass need to be established for both the thermo-chemical as well as biotransformation processes to convert it into fuels & value added chemicals.

"The chemical properties of the lignocellulosics make them a substrate of enormous biotechnological value. Plant biomass comprising lignin, hemicellulose and cellulose is one of the foreseeable renewable resources on the planet. .... Also it is apparent that integrated bio-refineries, analogous to petroleum refineries, would generate chemicals from the biomass thus moving the world towards a carbohydrate based economy.

"Cost of enzymes which is one of the critical factors for the process to become economical, needs substantial reduction. Major breakthroughs are needed to reduce the cost of producing cellulases, hemicellulases and to bring improvements in their activity and efficiency. A concerted effort in understanding the basic physiology of cellulolytic microbes and the utilization of this knowledge coupled with engineering principles is imperative to achieve a better processing and complete utilization of the plant biomass.

"The complete bioconversion of hemicellulose to sugar monomers is considered to be the main hurdle in the commercial success of bioconversion processes from the technical as well as the economic point of view.........".

"Lignocellulose biotechnology from investment perspective could be an attractive option for developing countries since its biodegradation could be achieved by solid-state fermentation, thus making such technology suitable for farms and small industrial plants without the need for large engineering infrastructure. The need for more efficient bio-separation processes leading to pure enzymes with less number of steps and lower costs needs to be developed towards commercialization..........."

Bioethanol

This chapter of the report has three specific sections: Ethanol from biomass, Bacterial ethanol production, and Bioconversion of cellulose/hemicellulose to ethanol.

Feedstock for bio-ethanol are sugar feed stock or grain based. The details are explained in the report. It is stated that technologies are well established. For a longer term horizon say 10 years or more the report emphasizes that lignocellulosic feed will have to be resorted to, when the demand exceeds the limits which can be served by grain feed stocks. The report lists the research challenges. It also states that "at the moment commercial production of ethanol from lignocellulose does not take place anywhere in the world. However, a lot of industrial related R&D are being implemented". Many of these are described in detail in the report. Lignocellulosic fed stocks are rice straw, sugar cane bagasse, wheat straw and cereal stovers. Their relative availability in India is listed in the report.

Very detailed scientific/ technical details are given for bacterial ethanol processes. Many are at laboratory stages or at pilot plant stages, results of which are giving an optimistic forecast.

Under the section on bioconversion of cellulose/ hemicellulose to ethanol, the following are covered.

- International scenario and current technology trends
- Indian perspectives

- Action plan
- Identification of a few projects
- R&D thrust areas and priorities
- Techno-economics
- International technology trends

Biochemicals

This chapter of TIFAC report deals with many chemicals. We give a few quotes from this chapter to get a glimpse of what is ahead. "Throughout the world intensive research has been initiated for developing chemicals from renewable feed stocks utilized in integrated bio refineries include, but are not limited to: grain such as corn, wheat sorghum, and barley; energy crops such as switch grass, miscanthus, willow and poplar; vegetable oils like castor, palm, soybean etc., and agricultural, forest, and industrial- residues such as bagasse, stover, straws, forest thinnings, sawdust and paper mill waste. The GRAIN study expects that the worldwide biomass supply can increase from a current 30 to 50 EJ upto 450 EJ per year in 2050 or 25 billion tonnes of dry biomass. The vision is to investigate new chemical strategies and develop new process for all partial replacement of fossil resources by renewable materials within the next 20 years..."

Bio-refineries are analogous to petroleum refineries in concept; however bio-refineries use biological matter (as opposed to petroleum or other fossil sources) to produce food, feed, materials, fuels, chemicals, and heat and power. Such bio-refineries would be capable of extracting from biomass, building blocks such as starch, hemicellulose, cellulose, lignin, proteins and lipids; and then converting them into a wide range of products deal with almost every possible aspect of our daily life.........

"Bio-refineries broadly divided into conventional bio-refinery, green bio-refinery, whole crop bio-refinery, lignocellulosic feed stock bio-refinery, two platform concept bio-refinery, thermo-chemical bio-refinery and marine bio-refinery. By producing multiple products, a bio-refinery can take advantage of the differences in biomass components and intermediates and maximize the value derived from the biomass feedstock. A bio-refinery might, for example, produce one or several low-volume, high-value chemical products and a low-value, high-volume liquid transportation fuel, while generating electricity and process heat for its own use and perhaps enough for sale of electricity. The high-value products enhance profitability, the high-volume fuel helps meet national energy needs, and the power production reduces costs and avoids greenhouse-gas emissions............."

"So far, the vast majority of bio-refineries are single process sites producing fuels and chemicals using first generation technologies. In the near future, more second generation processes are likely to become established. A number of technical hurdles have already been identified and addressed for the conversion of lignocellulosic material to ethanol, thus opening the door to the synthesis of other molecules".

We do not intend to cover other details or other chapters listed before.

The above quoted glimpses themselves would have given the readers an idea as to how much opportunities are ahead in the use of biomasses that is, from the plants which grow in soil.

Indian agriculture can transform itself into a mega industry with 10-15 years!

It is also good to bring to the attention of the readers as to what China is planning. There is an excellent report titled "Science & Technology in China: A Roadmap to 2050" brought out by Chinese Academy of Sciences with the Editor-in-Chief being Yongxiangby Lu (2010). The thrust of the whole report is about modernization of China, and it says that it calls for a New S&T revolution.

It gives a pride of place to agriculture. We quote a brief summary:

"Demand for agricultural products will rise substantially in China. There will be an increasing demand for food and fibre. Consumptions of dairy and fishery products will be triple in the next 5 decades. China's agricultural development will face both great opportunities and enormous challenges. The major development opportunities include expanding agricultural market, increasing comparative advantage of agriculture, improving agricultural structure, and agricultural S&T development. Meantime, serious security threats in arable land, water resources and eco-environment, the conflict between small farming and agricultural modernization, trade-off between food safety and rising demand for agricultural multi-functions, and impacts of global climate changes on agriculture are major challenges which China will have to face in the coming decades. Therefore, we must employ S&T innovation to build up the system of ecological and high-value agriculture and biological industry. Efforts must be done to facilitate smooth transformation of China's agriculture toward high yield, high quality, high efficiency, and eco-friendly agriculture, which will also improve China's food safety".

The readers need to closely examine the compression of words which summarise the detailed targets given later in the report for agriculture and bio-industry for 2020, 2030 and beyond.

- Consumption, tripling.
- Expanding markets, comparative advantages of agriculture
- But also serious security threats
- Conflict between small farming and agricultural modernisation.

- **Trade-off between food safety and rising demand for agricultural multi-functions.**

The report uses the word bio industry also along with agriculture.

Chinese have arrived at a plan. They are serious about implementing detailed projects and **also for learning by doing.** They do not stop at laboratories or in pilot plants. The targets given inside are in terms of all population; China's global share etc.

What is Ahead for India?

My main purpose in writing this article is to alert Indian scientists, technologists, engineers and technicians (in various fields, not limited to agriculture alone) academicians, business persons, public policy thinkers, policy makers, media persons, administrators and many others, to look at Indian agriculture beyond the usual narrow band coloured glasses of poverty, crises, subsidy, poor-monsoon-scare etc.,

Mother Earth has blessed the Indian geography with rich arable soil, bright sunshine and water resources. In many ways India is better than other countries, in terms of such endowments related to land. No wonder India shares 17% biodiversity of the world with only 2% of geographic area.

Are we going to waste the new opportunities which are opening up? Are we going to get lost with our usual myopia and muddle along? And later wake up and whine, when others have moved ahead? Let me point out a reality: We cannot wait upto 2020 or 2025, when many of these will become realities in China, USA, Brazil and many other countries and start talking about "catching up".

There is a natural competitive and comparative advantage for India in agriculture and bio-industry. It is very easy to skill our people to take up the new forms of economic activities based on land (soil) and plants. It will also absorb many ordinary Indians in productive employment. But the actions have to be based strongly on scientific, technological and economic foundations, and not on sudden raising of emotional one-point slogans. We have to learn from the experiences of the world, which are well covered in the TIFAC report and the realistic and well researched aspirations and plans of China, given in the quoted report.

❐

13
Invasion Biology

In this series, which is mainly focused on the future, we tried to bring out articles which provide a few strategic perspectives emerging in the world and possible alternatives for India; we also try to discuss some elements of newly emerging scientific areas with the hope that some of our scientific and academic institutions can capture the opportunity to be researching in those fields early enough before they become mature elsewhere in the world. One book has come out. The remaining thirty articles with some editing are included in this book.

In the post independent India, though funding for science and technology, and for academic institutions grew manifold, our academics and researchers have been timid. Mostly they did not venture to enter into untrodden paths. It is understandable that in engineering and technology, country as a whole was laggard; we imported machines, technologies and engaged management consultants from abroad to open industries in the public and private sector. Even the mighty organizations of atomic energy, space and defence research, had to follow the leaders in USA, Europe and Russia to acquire internal (technology) developmental capabilities.

But such a constraint, in principle, did not exist for basic research. However the scientists and academics were concerned about getting (individual) recognitions abroad. Most of them thus were followers of what was done by the foreign scientists.

They bought similar or same equipment, used by the foreign (developed country) scientists. The experimental methodology was same or similar to what was used by the foreign scientists who were pioneers or leaders in the emerging area of research. Even the models for data analysis were (are) the same. Of course, it helps international comparisons; also the Indian scientists and academics would get a chance to publish in international journals, adding "prestige" to them. Nowadays promotions of Indian academics are dependent upon the publications in foreign "prestigious" journals and citation index. The natural corollary is that Indian scientists and academics will crowd such areas of research where they can get better citation in foreign journals. People also resort to collaboration with foreign scientists so that the citation index goes up.

Thus the priorities of the scientific research (basic or applied) are not set on Indian needs or Indian context. No wonder there is a continuing divergence between our academia/scientific community and the actual end users in agriculture, industry and services sectors. This is so even in the Indian agricultural research; extension services are frowned upon as being inferior to basic research. But that basic research, which is encouraged, has little to do with the context of Indian farmers, trades and agro-business community. It is against this scenario, we tried to write about "Soil Biology" which, if done well, has relevance for soil conservation and soil reclamation. Also we took care, not to choose topics which will be totally India-unique, possibly new areas of original break through by Indian scientist/academics, as it may be rejected outright, in the background of the mind sets and governance system we have for S&T and academic activities. "Soil Biology" is now an accepted area of established scientists in developed countries; for over a decade or more. Indians can follow and also strike new paths, if they try.

The topic of this chapter is again emerging as an "accepted area" of research and academics. For India, it has relevance in the context of its huge biodiversity and the concerns of ecological balance. If pursued well, some Indian scientists/academics can reach leadership position in a few niche areas under this broad topic.

When I was searching for an article on Chinese anti-missile and strike capabilities over the ocean from Chinese eastern borders to the Pacific Ocean in the Time magazine (July 28, 2014), I came across an article "Invasive Species". It is an interesting article. That led me to study several other articles and I present here a glimpse of what I understood and what, I feel, can be done in India.

The fact that an article has appeared in Time magazine, which does not normally specialize in science and technology articles, indicates the strategic relevance for USA, in various walks of life.

Invasion Biology

There is a recent book by this title by Mark A Davis, Oxford University Press. The book is described as "the first synthetic, single authored overview of the field since Williamson's classic 1996 book; identifies current areas of controversy as well as future research directions; carefully balanced to avoid distinct taxonomic, ecosystem, and geographic (both investigator and species) biases; ideal graduate course material."...... "The book addresses a wide range of invasive species (including protists, invertebrates, fungi and plants) which have been studied in marine, fresh water, and terrestrial environments throughout the world by investigators equally diverse in their origins".

Above all, I find this description most important: "The book connects science with application by describing the health, economic, and ecological impacts of invasive species as well as the variety of management strategies developed to mitigate harmful impacts."

Suchan emerging field and a complex one, has passionate critics with differing (often opposite view points). In the web search one can find a review of another book "Invasion Biology: Critique of a Pseudoscience" by David Theodoropoulos (DT) Avvar books. The review is titled "Another Kind of Genocide". It states that DT is a conservationist and founder of an excellent resource for seeds of multifunctional plants. DT goes beyond "an examination of invasive-plant science (more properly the lack of it) and also explores the psychological, political, and cultural reasons behind our eagerness to hate certain species." DT's analysis and critique raise fundamental questions about "the idea that certain organisms belong in certain places and others don't". Why to label species as "nefarious", "stealing", "storm trooper plants" and "intruders"?

The review of the DT's book states:

"DT reviews the claims of the invasion biologists: that we are in an invasive-species crises: that humans are moving species faster and farther than nature even could; that many species have become extinct solely due to exotics; that invasion corrodes ecosystem processes, lowers diversity and is destroying our wild lands. Admirably, he goes to the original sources cited by the doomsayers, and shows that the data do not support single one of these claims".

I have quoted from the review of DT's book by Toby Hemenway (http://www.patternliteracy.com/201-another-kind-of-genocide, (2005) rather extensively because the "Invasion Biology" without such a scientific critique is more appealing and can easily bias a reader and make them xenophobic of anything different. In fact, the Time magazine title is very tendentious: "Invasive Species" with a picture of a python covering two pages. Of course, it is a very informative article, giving considerable information about examples form USA: a python native to Southeast Asia, now spreading in USA; African giant snail;

Cuban tree frog; snakehead fish from Africa and Asia; emerald ash borers from Asia and eastern Russia; Asian carp and several more. They come through cargo carrying ships or pet animals. If one does not read many other articles (some of which I had cited), one could be left with a doomsday feeling.

Numerous Concepts of Invasion Biology

From the above cited examples and quotes, one has to conclude that this is an area that needs to be studied scientifically, without rushing to knee-jerk conclusions of "extermination", "quarantine" etc., Nature is complex. Biology is vast and elusive.

As the reviewer Toby Hemenway states "Theodoropoulous seems to have modeled his book on The Origin of Species, wherein Darwin assembled an overwhelming number of examples to support his view and in effect crushed his enemies by sheer weight". One important point which emerges is that we need to have the Darwinian patience to assemble facts before rushing to conclusions, one way or another. In the process humanity may become much wiser and perhaps be rewarded with the knowledge gained. Perhaps DT's conclusion may be right "Nature creatively mingles both native and exotic without prejudice".

Going a little back in time, I am reminded of the raging panic during the mid-1970's and early 1980's against parthenium in Bangalore. It is from a seed which had apparently come along with Mexican or American wheat imported by India. There were many newspaper articles, other "instant" "scientific" papers in Indian journals. The allergy from parthenium was painted in gory ways; some reports said that women were not affected as much as men. I had not gone into the sources of the studies (if any) and data sample. I lived in Bangalore then. General hysteria was that with this dangerous weed, most Bangaloreans would suffer from asthma and allergic diseases. As usual, the burden of removing them was put on the government. Decades have passed by. The parthenium plants are everywhere in Bangalore, happily settling down with other plants. There is no hue and cry now. There does not seem to be any increase in allergies. Now Bangalore's problems are the traffic congestion, garbage, fear of dengue because of mosquitoes etc.

Looking back further into the millennia, if we tell that chillies (green and red), potato, brinjal, tomato etc., are not original Indian plants and they were brought into India as exotic plant, one may get a shock. Now they are all over India. Nobody will believe that chillies, which are the soul of Indian cuisine are not indigenous. Perhaps even tamarind came as "invader species" brought by humans.

When I was a young boy, I have seen (and eaten) the special food cooked in Brahmin households for the death-anniversary function (shraarddham). The disallowed items for that day's cooking are: toor dhal, chilli, potato, brinjal and tomato. The things allowed are: Moong dhal, black pepper (instead of chilli); cluster beans (cheeniavaraikkai or kotthuavaraikkai in Tamil); banana all varieties-raw to fruit, urad dhal (for vada), and a few more. At that time, I did not know the reason; the household persons did not know. Later when I learnt about items brought into India such aschillies, I could guess how the tradition has come about. Only the "original" ones or "native ones" might have been prescribed in some written rules, in a "sastra" or other such equivalent ones.

I have particularly narrated these items because India with its rich traditions in different parts of India including that of tribal people, those who live in remote places etc., would have such taboos or preferred items of food; or preferred items to be used for special functions. If scientifically and systematically analysed (also correlate to ingredients of Ayurveda, Siddha, Unani, Tibetan and Folk medicines) a good scientific picture may emerge as to how the alien species started replacing/coexisting with other native species. How much of it was due to biological success, how much of it was due human (cultural) selection? How much of it was due to commercial considerations such as better productivity/yield?

Indian language and literature, archaeological findings, anthropological studies and use of modern genetic methods may bring to light a whole new set of evolutionary processes. These may also provide us with clues to handle the newer species which would come into India as microbes, insects, fungi plants, birds and mammals.

After this diversion, let us look at the gist of a recent paper with the title as given for this section (we have used the title of that review article by Jannike Falk-Peterson, Thomas Bohn, and Odd Terje Sandlund, in Biological Invasions (2006): 1409-1424, Springer 2006. Author for correspondence (e-mail.jarnikefa@ hotmail.com). The abstract well defines the content and scope of the review paper. "The study of biological invasions has triggered the production of diversity of concepts. The terminology has, however, often been applied inconsistently and inaccurately. The article lists and assesses the most commonly used terms and concepts in invasion ecology. In each case, the most coherent definition and use is suggested".

A figure gives a schematic overview over the domains and main elements of 'classical biology'; versus "invasion biology". In the domain of "classical biology" there is natural distribution and natural migration. These are within the natural boundary, which has to be judged over centuries and millennia. But with the spread of human vectors with ships, surface transport and also air transport (or human and also goods and material), the native from its natural system gets 'introduced' to a receiving ecosystem (which is non-native). Then there is 'naturalisation' and also 'invasion'.

There are tables in the review paper giving terms and proposed definitions. The terms are: Endemic, non-native/alien/

adventive/exotic/foreign/introduced/non-indigenous/novel, Introduction, etc. etc. Then details on each of these terminology/ concepts are given.

In the conclusion the paper states "The growing scientific field of invasion biology has suffered from its often confusing use of terms. Synonyms have been too many.................."

"Conceptual confusion, value laden terms, and taxonomy-based barriers within the field of invasion biology are all unnecessary............. We encourage all invasion biologists to tighten up their use of terms and concepts for a common goal of clarity........."

The paper gives in the appendix, an overview of the 145 selected concepts and definitions used in the literature, sorted to 10 main categories. In addition there are copious references running to about two pages.

We stop here without quoting from several other excellent literature on the subject, we have studied.

We could not do an exhaustive search about Indian work in the field. It appears to be less. While doing so, we found the vision 2030 document by the Directorate of Weed Science Research, Jabalpur (www.nrcws.org). Though it contains elements of invasion biology, its focus is different – to solve the problems of weeds. But within that also there is a scope for Invasion Biology science per se.

To Sum Up

There is a new emerging area of biology in which it is possible for Indian scientists/academics to make a good international presence. It has relevance for ecology, biodiversity and perhaps several other applications in future. Our idea was not so much on such immediate benefits, but to find areas in which Indian science can make original contributions to the corpus of world knowledge.

If our scientists in addition to pursuing "Invasion Biology" in the narrowly defined terms (which is essential to make an international presence), also attempt multi-disciplinary studies involving archaeology, cultural practices like food, use of preferred plants, birds and animals, study of early literature in Indian languages (ranging from a few millennia before), and folk literature, many new knowledge domains may be opened up. Let Charles Darwin be the ideal in terms of collection of samples, data and putting them together. Modern genetic studies of these older plants, fossils etc., may also provide deeper insights.

We have avoided discussion about another form of 'invasion biology' (which is predominantly due to the fast, global human travel, transport of food materials etc.,) which are global pandemics, latest being Ebola, and many others during the past decades like H1N1, bird flu, SARS etc. They need to be taken on a war-footing, though we may not understand most of the science behind their spread.

But the invasion biology of other organisms like plants, fishes, fungi, insects, animals, birds etc., requires a more careful systematic study, though even in that field, there could be some scare-mongers, as pointed by David Theodoropolous. I am confident that the processes of science are reasonably robust to take care of such occasional xenophobia.

Let me appeal to the Indian scientists to pursue more of emerging biological sciences, which may not all require huge investment of capital infrastructure. They have the potential to give the Indian scientists spread all over India, an opportunity to do some original first rate science.

❐

14

Earth with 10 Billion People

Most of us in India do not consider population as a major issue at all. A very small minority opinion is that "control population growth and India's problems will all be solved!"In this chapter we are not considering family planning or population control on a global or national level.

We are attempting to present here some glimpses of an excellent recent study titled "Can earth's and society's systems meet the needs of 10 Billion People? Summary of a workshop" by National Research Council (2014) published by The National Academy Press, Washington D.C. www.nap.edu.

In India it is very rare to have a systematically conducted study on emerging future scenario. Even the well-established universities, academic institutions and various national academies of science, engineering etc., do not get together several scholars and practitioners, to apply various models to future scenarios nor to conduct detailed surveys. We had left all these tasks to the Planning Commission. Most of the Planning Commission reports turn out to be the listing of "desires" of various government departments and then "bringing them within the budgetary constraints" of five year time frame. Thus these projections over a period have lost their authenticity in terms of modeling trends globally and nationally.

In the developed countries like USA, Europe, and Japan, there are many think-tanks which do this job regularly in various fields. We have given a glimpse of one such study for 2030 in chapter 26 of the earlier book. The above quoted NRC workshop study report gives an account of various models projecting to 2050 and in some cases even to 2100. They are not science fictions or economic or social fictions. They are serious attempts to look at several possibilities projecting the trends of the past and also modeling the current status. The authors point out the uncertainties and constraints as well. On the whole a reader of the report can get good clues about facing the new challenges that are ahead of us as humanity and also the new opportunities.

It is also possible for individual nations, especially the larger sized ones to use the approaches given in the NRC report, order to project their own future options and the corresponding steps they need to take in the near – term future. In fact the developed countries do such studies only to get some idea for themselves about global strategies and also naturally to maintain their leadership roles.

The Question Posed

Right at the beginning of the report the first paragraphs states: "The Earth's population, currently 7.2 billion, is expected to rise at a rapid rate over the next 40 years. Current projections state that the earth will need to support 9.6 billion people by the year 2050, a figure that climbs to nearly 11 billion by 2100 (United Nations 2012). At the same time, most people envision a future of the earth with a greater average standard of living than currently exists – and as a result, greater consumption of planetary resources. How can this population growth be achieved in a manner that is sustainable from an economic, social, and environmental perspective?"

The workshop organized by the NRC "was designed as a venue to explore issues, not to reach consensus". It is an important approach we need to adopt. The issues involved are complex and some of them may have strong ideological and/or political overtones. Also when a workshop is held in India, the expectation in India is that they will come up with consensus solutions.

The workshop was organized into six sessions focusing on the following concepts:

- The Human-Earth System
- Challenges to the Earth System: Character and Magnitude of the

Challenges in 2050

- Challenges to the Earth System: Consequences for the Earth System.
- Extreme events
- Resource Distribution and Global Inequality
- Interaction between Earth and Social Systems.

Recurring Themes

The report also lists in detail the recurring themes and their sub-elements. We present them here in a compressed form:

(1) Demographic variables that influence sustainability: The sub elements discussed in detail are: Education, Fertility and Aging, Urbanisation and Migration.

One of the experts (Bongaarts) in his presentation has stated that "fertility rates lower than 1.7 or so births per woman result in a demographic imbalance, with an aging population that puts stress on the economy and society. This imbalance is among the biggest challenges facing United States and Europe in the next two or three decades.

Though the report does not specifically address India, we need to think about India, while reading the report. India could reach 1.7 to 1.8 billion population level by 2050. Also the rate of growth of population may fall; its aging population is going to be quite high. Much of the aging population (if our current trends of employment in unorganized sector without any form of social security cover continue) will be without financial resources. Also urbanization will drive families to be more and more nuclear. The population of senior citizens currently about 100 million may rise above 200 million.

Again there are further sub-texts as issues of inequalities are discussed in the report. India is not just one homogeneous society. The rich-upper middle-middle-lower middle-bulky base divides will continue. With faster economic growth the size of the bulky base may come down in three decades and current levels of poverty would definitely come down. But differences between the divides may increase.

Migration within India and to other countries may increase. Also India's neighbouring countries may contribute to migration from outside. Many upper middle-middle class Indians may like to settle down in USA and Europe and perhaps in Japan if it relaxes the immigration policy to fill their gaps for lack of youthful workforce. The experts who presented at the Workshop state that presently about 3% of the population migrates but about 8% would like to do so. We could not get India's specific number. But Indians would like to migrate seeing the past and present trends. Also the well-being of People of Indian Origin (PIO)'s living abroad will be a great attraction for more Indians to go out.

But let us also not forget that many persons in Europe are getting wary of their multi – cultural experiments of the past. Spread of global networks of terrorism is one of the reasons. Present (2016) reactions to immigration to Europe may become policies.

(2) Economic and Policy variables that Influence Sustainability

The sub-elements are: economic incentives; food: costs and consumption habits; land use; technological change; we don't list all of these here.

One of the presenters Rouse has explained that "the adoption of an innovative technology is a highly selective process, and those who embrace a technology or innovation first will experience it very differently from subsequent adopters. While adoption of a technology usually also leads to increase in inequality, because early adopters tend to receive the greatest benefit". He further noted that "near-term technology tends to benefit those who least need the improvement".

I see that there are two major messages for us. In India we tend to follow a "wait-and-watch" policy. We take decisions to change very slowly, usually about 7-10 years later than when it is needed. We have addressed this Indian time-rate-interaction in many chapters in this series. Look at adoption of GM crops or agricultural reforms or manufacturing policies or monsoon-proofing-of-agriculture, etc. The later we do, much lesser will be the benefits: early birds would have caught the fatty worms!

Near- term technology tends to benefit those who need the least: obsession with quarterly results tend to crowd innovations or technological change towards near-term goals. In public perception, flashy goals look better. Creation of many institutions for ordinary skills does not look as attractive as opening of more IIT's or investment in high sounding technology research.

India has to chase faster growth in the next three to four decades (especially because of serious loss of pace during the past few years). It is important to look at the composition of the change; not enough to chase 8%, 9% etc. without understanding how the benefits are demographically balanced.

(3) Suitable Metrics: Sub-elements are overall human well-being; indicators of environmental change, including tipping points and boundaries; impact of climate change on humans, demographic changes, return on investment.

These are highly technical with many models presented and discussed.

The other themes include: **Carrying Capacity.** A general agreement appears to be that since human beings have acquired capability to manipulate various natural limitations on animals, "the carrying capacity has been "largely abandoned" in the social and policy scenes." Other two **themes** are:

- **Integrating Social and Natural Scenes.**
- **Structuring a Research agenda.**

The introductory chapter ends with a bullet point:

"Consider not just the science, but also the policy, including ideas for implementation".

Glimpses of Discussions

The subsequent chapters discuss the papers, presentations and the question – answer sessions of these themes. The rates of population growth will be the highest in Sub-Saharan Africa (SSA) and the Middle East. In SSA, 70% of the population is under 25 years. In the developed world, population under 25 years is 30%. In between, is much of Asia reaping temporary "demographic dividend" of rising proportion of workers – but also in three decades, they will be aging. Then the immediate effects on consumption, population etc., are discussed.

Under urbanisation, the presenter Marcotullio made five statements:

"(i) Urbanisation is a mega trend for the 21st century.

(ii) There are challenges associated with environmental sustainability and urbanisation;

(iii) There are many opportunities to gain efficiencies;

(iv) Despite these opportunities the scale of issue is overwhelming.

(v) Humans have the potential to create biophilic, resilient, liveable cities in the future".

There are many details. We would like to point out that if done carefully, India can convert the opportunities to its advantage domestically and globally. While the report is about global sustainability, there are also regional differences. Each nation would like to derive benefits. Our emphasis is that within the overall context of earth-system, India and Indians can derive benefits, if they plan and execute well. Of the 100 SMART cities announced by the Hon'ble Prime Minister, let us hope that we would also make them into **biophilic, resilient, and liveable cities** – the main overarching theme of this article series: Living with Nature in the Modern Forms.

One interesting issue which is pointed out will be relevant for India as well when we consider the regional and social variations in India.

"The child-bearing and child rearing ages can take up the entire life of a typical African woman: She can have children from ages 18-40 and would be 60 before the younger children left home. A German woman, by contrast, more likely has two children very close together in age. She would be absent from the workforce for only a few years.

Indian situation has many differences. In the poorer bulky base of the society, she may have more children, still work in subsistence agriculture and other manual tasks, with little income. Her counterpart in the middle-class may have one or two children. But many of them are absent from workforce due to various other socio-economic compulsions including taking care of old age parents and in-laws. On the whole urbanization and education does not necessarily draw women more into the work force. On the other side keeping her in rural areas with poor incomes and lesser education, affects the future generation of work force (including the boys she rears) as the educational and social opportunities around rural areas are poor.

We have addressed the skew in the women-work-force data in India in an earlier chapter. It is going to be a critical issue especially when the aging population goes up in number and the available young woman population is deprived of work-participation (especially in higher skilled jobs).

Discussions on consequences to the earth system give many important points. The experts have also used the words "ecosystem services" to describe the benefits the eco system gives to people.

Two challenges identified by Polasky are:

(i) "economic development to alleviate poverty and malnourishment and

(ii) conservation of natural capital to maintain biodiversity and ensure that humanity fits into the earth's system".

Polasky further stated that "to support 10 billion people in 2050, the world GDP would need to increase by a factor of eight."

After detailed discussions of these (apparently) contradictory needs, he further says about eight-fold increase of GDP that "simply scaling up business as usual" cannot produce enough for a population of 10 billion." He called for improved technology (better output per unit input), as well as improved institutions and incentives to increase output of goods without a negative impact on natural capital".

We skip many other chapters and discuss India specific issues, based on my understanding of the reading of the whole report.

Now Looking at India

For India, the above observations are important. It has much less geographic area for its population compared to USA, China and Brazil etc. It is going to have a huge population about 1.7 to 1.8 billion (about 17 to 18% of the world population). A large number of them are consuming very little, by global standards, even sub-Saharan standards in many cases.

Just one example from a brief report (in figures) in Times of India August 29, 2014 about world consumption of all vehicles and passenger cars, given in terms of number per 1000 people.

Region	All vehicles per 1000 people (Not including two wheelers)	Passenger cars per 1000 people including taxies.
Advanced economies	1013	460
Emerging and developing	590	275
Europe (excluding USSR)		
Former Soviet Republics	389	177
Arab countries	372	164
Emerging and developing Asia	320	141
(Excluding SAARC)		
Sub Saharan Africa	111	44
SAARC economies	48	18
Number of cars per 1000 people		
Sri Lanka	20	
Pakistan	16	
India	16	

There is a huge hunger for modern goods and services from the Bulky Base, Lower Middle Class and Middle Class. The growing Upper Middle Class desire "developed country standard" for all their consumption *(Nothing less)* with faster economic growth UMC will grow larger in size; with ICT penetration their voices in social media will be much larger (as the current trends are) and therefore in electronic media.

The technological, institutional, governance, political challenges for India are much more and real. One cannot go on with severe inequalities either. Consumption has to go up, if standards of living have to go up. We may have to develop different standards; over-zealous advertisements aimed at UMC's alone need to be toned down, giving equal importance to other "standards" of products used by LMCS's, BB's etc. which may be such as to place lower "Consumption Handprints" from the natural capital.

Since we are late entrants to the game of faster economic growth, we may not be able to follow the foot prints of S. Korea and also that of China, which went through recently a severe high-energy consumption, high-pollution generating phase. China is quickly correcting these problems.

India will be expected to follow pollution and emission norms. There may be new conditions from World Trade Organisation (WTO). Let us remember the tough period we have with Food Security Provisions. The developed countries, and those of the countries which have acquired reasonable standards of living, would join together. Even our own UMC's and MC's would like to have almost the same pollution and emission standards of the West. If other Indians cannot afford to pay, why should they bother? We have a tradition of government subsidy.

But these issues relating to protecting natural capital cannot be solved by individual subsidies, though certain incentives to protect nature will be useful.

There is a statement in the report, almost at its very end; "Berkman responded that many countries (notably India) have neglected those with the lowest socio-economic status in favour of supporting the rising middle class. This is a country level policy issue..........."

We, as a country, need to go beyond the usual politics of pressure-groups especially the more vocal ones. At the same time as it is given in many parts of report discussing inequality, it is impossible to have a near-total equality even in the horizon of three to four decades.

Enyegene pointed out that "some inequality was suggested as fundamentally good to spur innovation and effort, although others might argue that inequality is on principle a negative characteristic........"

Again it is pointed out that most economists will point out Gini Coefficients for inequality measure. But in that process they equate rich American with rich African. One interesting information is worth noting in this context. "........... the mean poverty line globally is $ 1.25 per day, but in United States the mean poverty line is $ 14-15 per day; this amount is equal to 75th percentile of world income distribution."

We should remember this gross historical inequality between developed nations and other nations. Countries like India had missed a lot during the colonial period and to be truthful also during many decades post-independence. India is emerging and more Indians are consuming better (earning better). Whether they will also consume the same amount or earn the same amount is an issue. Realism shows that it may be a near impossible target for 2050, unless the consumption levels for all is severely brought down. That will affect the economic growth, impacting the poor more. Also national or global natural capital cannot be saved.

So Indians have before them a complex set of problems to solve. But in the overall, Indians have been a tolerant and resilient people.

In the emerging picture of 10 Billion people on earth, we cannot decide how all others will live. But Indians can choose the path for the 1.7 to 1.8 billion Indians, around 2050. They can, not only, help themselves well but also help many other people of the world who may have problems in food production (especially if climate change impacts take place), in health management etc. In that process, Indians can also have new economic and social opportunities. As we have described in many of the chapters India and Indians can make a better future for themselves and also for a number of people of the world.

And in the process, live more harmoniously with nature.

❐

15
Swaccha Bhaarat

Even when I started writing this article, I was still through the pangs of the great loss of a sage and Karma Yogi from our midst on October 2014. Arulchelvar Mahalingam. He will still continue to guide us, if we try to listen. On October 02, 2014, the Prime Minister launched a major movement: Swaccha Bharat (SB). The actual mission or movement is much bigger than broom sticks and latrines.

Cleanliness as we understand in the modern world is a very sophisticated combination of hygiene and aesthetics. At the early time of human evolution, cleanliness had low priority. The main emphasis was survival of individual humans and propagation of the race. As human prosperity (even for a small percentage) and collective security improved, cleanliness became an integral part of life. The remains of Mohanjo Daro, Harappa, Lothal etc., bear an example the progress of habitat planning, urban planning etc.

The religions and literary writings emphasized on many aspects of cleanliness. Many rituals absorbed the elements of cleanliness.

It almost became axiomatic in all human societies that the civilized person will necessarily observe high level of cleanliness. With the progress of modern science, engineering, technology and medicine, the entire concept of cleanliness (swacchatha) took new dimensions. It is very much coupled with aesthetics as well. That is the reason why the Prime Minister while listing the importance of Swaccha Bhaarat (SB), points out that it is an essential condition to attract more tourists. Every Indian is aware of the linkage between cleanliness and health, though the depth of their scientific understating about cleanliness may vary.

Equally well understood is the linkage of water with purity: "Aapa Punanthu Prithvim" says Vedas; (water purifies the earth). That is the reason that most of the Indians will at least symbolically clean the hands with water before eating. People enter many places of worship after symbolically cleaning their legs and hands.

But would these symbolisms suffice? Most of the water bodies (ponds, lakes, etc.) in and around temples are dirty. The Gangaa is highly polluted. Worst still is that many Indians have to live close to their own excreta, urine and feces and other wastes which they generate. Many towns and cities are littered, in addition to garbage, by animal excreta of cattle and dogs.

One Point Slogans

Since the Prime Minister, with the ability to give attention to details in addition to giving a Vision (and catchy slogans), is directly involved in this SB Movement (SBM), there is a lot of hope that we will not get into the usual activist and bureaucratic traps of one or two point projects, quantitative targets, inadequate funds, subsidies and above all forgetting the main elements of the complex processes of SBM and their linkages.

This article is meant to point out a few of these linkages, so that in the actual execution of SBM, Indians derive benefit: better health for themselves and others; better aesthetics around their habitat, work places and public places they share; an excellent attraction for outsiders who come to India to do business or enjoy its beauty or research on the heritage of a great civilization.

And above all we will, through proper execution of SBM, leave for the future Indians (yet to be born and those infants who are too young now to understand), a great India which is clean and beautiful (which was dirty and filthy only till a recent past) and a confident India which has forgotten those putrid conditions of existence.

In order to achieve this state for India and Indians, we should avoid many pitfalls and land mines which exist amidst us.

Bureaucratic Traps

In a recent article by Santhosh Malhotra (6-10-2014 Times of India) about Swaccha Bhaarat, a number of central Government programmes are listed. The Central Rural Sanitation Programme (1986-99), Total Sanitation Campaign (TSC)1999-2012, and the ongoing Nirmal Bharat Abhiyan (NBA) 2012-22. The author states that all have failed. TSC was started by the earlier NDA government to complete it by target year 2012 to achieve Nirmal Bharat. The author says that instead with a population increase and with toilet coverage increasing at only 1% point per year, there were 8 million more households defecating in the open in 2011 than in 2001 (census data)". The author further continues "share of rural households with toilets increased from 21% to 31% between the two censuses. At this rate of improvement, it will take nearly 70 more years before India becomes OD free" (OD-Open Defecation).

The way many such socio-economic programmes are designed (I am familiar with these since late 1980's) the bureaucracy reduces a complex problem into a set of targets to be completed by the State governments and release grants according to a pre-set (so called well-coordinated with state level bureaucracies) project format. Yearly completion reports and fund utilization certificates are received and filed. There may be some "monitoring committee" reports with little value. Therefore on the whole in many cases the government auditors cannot find many faults, except for some nit-picking on issue of tenders; project escalation etc.

One well respected top bureaucrat had told me in the late 1990's that many of these programmes are gurgitated and re-gurgitated. In addition to the above referred central programmes, there may be many state government level programmes with fund released to the panchayats and municipalities. Again all these unfinished projects may be brought up again and States may ask that the SBM funds may be given to these "on-going" programmes. There may be compulsions of federal set up (and politics) to accept these.

So far we have discussed about progammes and funding of them under various names for the past three decades. If through this bureaucratic maze, political apathy, inefficient government machinery and above all the omnipresent "leakages of funds" at various levels, some latrines are built up a big question is: are they serving people? (be they in their homes or built in public places like slums). Santosh Melhotra states in his article that "research evidence is that people defecate in the open because they do not see a reason to change their centuries-old behaviour". He also estimates that "perhaps four-fifths of the expenditure was creamed off in corruption and wasted. Disbursement was taken as a proxy for construction."

I have worked for a DST project called "Good Morning Bombay" by us, in 1989. The non-utilisation of toilets built by government or subsidized by government is not just because centuries-old behaviour alone. There are other pressing present day realities.

During my own life time from mid-1940's I had used the common latrines in Bombay Chawls (also later during 1957-1964). For about 4 years, did OD in the Tamil Nadu village (1949-1953) and used dry latrine in Palayamkottai inside the house (1953-1957). Only after 1964, we had a proper modern toilet for individual house usage.

OD as I did in the village open fields, was better than the dry latrine (cleaned once-a-day in the morning by a woman worker). The chawl like sharing a modern latrine was most atrocious, especially because of severe water shortage, technically they were Indian style and had a septic tank. A few lucky one's (we all use to compete) who can enter immediately after the first cleaning in the morning may have a clean latrine to defecate. Then due to water-shortage, it was a hell-hole. I used to get infected very often with a type of eczema in the thighs which I never had in OD days!

I have described my experience in some detail because I find that the question of defecation is often reduced to a toilet block. **Does it have water in abundance to clean not just once a day, but after every use?** Obviously it need not be water that has potable quality! Our drinking water mission is still to achieve the results of giving clean drinking water to all Indians! Even most of the regular municipal water supplies are so poor that the middle class, upper middle class and of course the rich and powerful use only bottled water even in five star hotels!

Linkages to Latrines

If latrines have to attract people to use them, then they should have:

- Good water supply to clean it after each use (and even before use if it is going to be a public latrine not limited to a family)
- Good ambience-with light and with no foul smells.
- A good (not choking) outlet for the excreta and water, through a large enough septic tank or through connection to sewage lines.
- Regular maintenance especially if it is a public toilet, with disinfectants poured often for cleaning.
- Also cockroaches should be killed as often as possible.

Many children and even elders may get scared if they are cock-roach infested. Many latrines in chawls of Bombay (now Mumbai) were full of cockroaches.

- Annual removal of the muck from the septic or more often as needed.

I have not suggested the use of human excreta for biogas generation, though technically feasible, as it will be resisted. Let us not load the project with many such goals.

Instead of counting numbers, in terms of disbursal of subsidies (which often are the sources of corruption) let the SBM concentrate on actual execution of **Operational Latrines** on the lines suggested above (taking care of all the linkages). Let it start at the village level, with proper construction of sewage lines and septic tanks at homes and/or public places **as the villagers choose.**

Let the corporate CSR funds be channelised for this; not for **Release and Forget** but for **Execution, Transfer and regular Monitoring.** Accounting for these may be difficult; some may over inflate the costs. **Can the government fix a comfortable figure so that there is no need to cheat?** Anything above that will not be paid. Anything under that is okay – let us leave it to them to enjoy!

There can be independent monitoring by local colleges. Let them send pictures and data. There could be some boys and girls who may try use this mode to create scare some persons, or use it for blackmail! Let the data therefore not be used by tax authorities to account for CSR expenditure. Then in a short period things will stabilise. Those who are bad in implementation can be caught by others, as such data is available. Those who do well need not be scared! **Truth will succeed if approval mechanisms, tax evaluations etc., are not tied up to the monitoring system.**

Latrines in Urban Slums

It has to have all the above linkages. But space available for latrines will be small, as land is prohibitively costly in urban areas. The Department of Science and Technology (DST), Government of India, under the guidance of the then Secretary Dr. Vasant Gowariker had executed a Latrine building project (pre-fab using ferro cement, with blocks of 15 or 30 each). It worked well during 1990-1992 in most of the 30 odd locations. It was on Build (at Government cost) and Transfer basis. Pay for use except women and children. The local communities operated them; there were some interesting sociological problems due to local mischief makers at some locations. The project was to be expanded by a public Trust, with top industrialists on Board. But alas! The 1993 blasts in Bombay disrupted any further work.

But similar work might have been done elsewhere. The idea is to use minimum of space, but taking care of all other items including aesthetics.

Swaccha Does Not End With Toilet Blocks!

Other items, to name a few ones, which destroy cleanliness in local areas are:

- Solid waste (Garbage disposal)
- Animal excreta
- Pollution of water bodies (not just rivers but ponds, lakes etc., in villages and towns)
- Waste water disposal
- Agricultural wastes in rural areas.
- Animal wastes in butcheries

About **Garbage in urban areas,** we had already devoted a full article in this series. It appears as the chapter 24 in the book Mission 21st Century (Living with Nature in the Modern World) published by Ocean Books Pvt. Ltd., (2014), the earlier book.

Another article was regarding Plastics (Chapter 2 of the present book) and it gives statistics of urban garbage all over India, amount of plastics etc.,

We won't repeat them here. **Urban Garbage Waste** can be converted to energy (solid fuel like coal, or as electricity). There are working plants in India. If there is a market, let them be converted to compost also. But the size of urban garbage is large and is increasing; to convert it into electricity is better.

We should take care to ensure that the projects do not get stymied by fads and fashions of single point activists such as:

- Segregate the garbage (We had argued how it is unrealistic for most of Indian households).
- Ban plastics bags! (We have argued with data against it!)
- Ban hawkers! (Let us not destroy livelihoods!)
- Ban rag picking! etc. (Same as above)

Let us take care not to disturb the livelihood of poor Indians until we find an alternate income generation jobs for them. Many of them can be effectively used in **Waste Collection, Disposal and Reuse, if systems are developed.** It can be a part of SBM mission, if industries are incentivised.

I found that after participating in SBM launch, many well-meaning persons have collected garbage and dry leaves in their local areas (through sweeping) and burn them locally. In Bengaluru as I had earlier written in my article quoted above, is indeed having such a system of "decentralized incineration" on roads, empty sites etc. It is not good in the long run, though for a quick look, things may look clean. It increases global warming and more importantly pollutes the local air. Some dump them quietly in a nearby water body, it adds to ground pollution, and loss of water.

Hence all urban areas even Tier III towns and large villages, have to adopt proper Solid Waste Usage Plants (SWUP). As India grows and gets prosperous, the garbage will grow. Look at the scheme to sell good packed food in railways. It is a great step for health. What happens to the packing which goes as garbage? Will they be thrown all across rail tracks? Better way is to collect them and dispose them in a number of railway junctions to be taken for solid waste usage plants. Big bus stations have to install good garbage collection facilities (never mind segregation). Standardized containers will help transport to SWUP.

Agricultural Wastes in Rural Areas are much bigger problems as farmers produce more. Most of the straw is burnt; instead these can be used to produce electricity or get bio oil. I learnt that IIT Ropar (Punjab) is working on a project with University of Austin. Hope they do it fast and implement all over Punjab, Haryana and western UP.

Animal excreta in towns, cities and metros: These are mostly due to cattle and free moving cows, on roads. For a Swaccha Bhaarat, we cannot allow this to happen. Such stray cattle may be confined to a separate place; if there are owners, they may be fined for letting them free.

On the upper end of society, many Indian who walk with their pet dogs in parks or lay outs let them defecate in the public places. I have seen it in the Jawaharlal Nehru Park in Delhi. The foreigners who come with their dogs keep plastic bags to collect the poop and carry with them! But Indians let them free on lawns, foot paths etc. One has to shame them and may be fine them. Should not SBM slogan reach such elite and middle class pet owners? As India develops more Indian households have started having pet dogs using the Western practices, but let them also imbibe in them the civic sense of the Westerners!

Animal Wastes Near Butcheries

We had once done a study in TIFAC (Technology Information, Forecasting and Assessment Council) about such animal wastes around late 1990's. The business from proper use of them can be around Rs. 3000 crores annually (now much more!); we were working with Khadi Gramodyog to implement commercial projects! But the Chairman changed; then project gets stranded. Even now, it is an area for SBM. It will improve health of Indians as well, as more people are going for non-vegetarian food as they get better incomes. The wastes other than mere leather and meat, can give good earning incomes too.

Water

Now we are left with two major items concerning water. In villages large amount of water is wasted in fields. They flow into water bodies and pollute them. Most village ponds and lakes (including temple lakes) have unacceptable levels of BOD (Biological Oxygen Demand). Temple ponds get spoiled because most of the milk etc. used for abhisekham goes into the ponds. People also feed fishes with prasaadam.

As we make more latrines in rural households the excreta and urine, and bathed water will find their way to the local ponds and lakes. If water disposal systems are not planned well, people will direct latrine outputs to a system which will finally go to a lake.

At the initiative of the then Chief Minister of Punjab in 2000-2001, we had done extensive preliminary work on cleaning of the lakes in Punjab; most of them were in bad shape. We were to start the work on a Waste Water Recycling Plant on a land donated by the villagers. Elections were declared; no new work! Priorities changed. So it goes! **This is the situation about most lakes and ponds in India. They are the life-line of Swaccha Bharat. There can be a waste water recycling plant for all villages in India. If necessary, they could be subsidized.**

The situation in urban areas is still terrible. We use ground water 365x24 hours and throw them in sewage. Some lucky ones get river water and do the same. Rain Water Harvesting can only work for a few days or weeks during monsoon months. So Waste Water Recycling Plants have to be there in many places in a town or city. About 2 million litres per day plant can be cost effective. Several such plants can be located in different places in a city. Some municipal corporations in India (Hyderabad, Bangalore are among the forerunners) have done it. But maintenance is the weak point of Indian psyche.

SBM should have Waste Water recycling and Reuse as the Prime Slogan.

"Aaapa Punantu Prithivim" – says Veda -, "Water purifies the Earth". If we purify used water, we purify the Earth as we prevent it from getting dirty through dirty sewage etc. We save the ground water also, from being drawn out. Ganga, Yamuna, and all the rivers will be free of dirt if the recycled water flows back into them! All our hotels, apartment complexes, industries, Lutyens Delhi – all should pay for recycling of water.

Water is the Mother of the Swacch Bharat Movement. Also people can avail water for sanitation purposes including many migrant workers, who do not have regular homes.

Let us achieve all round cleanliness in India: it is achievable in a decade, if we approach the issues holistically seeing the linkages and avoiding "attractive booby traps of files and single point slogans".

It is a great movement, started on Oct. 02, in the memory of Bapu! Let us achieve it sooner than later. "Kaalkarey so aajkar" said Kabir.

❐

16
Education: The Key

We have arrived at the fiftieth article: A regular reader of this series told me that fifty is a milestone and I should choose the topic very carefully. I was deeply thinking within myself as to what is the key for the modern world and India. I was also surveying Arulchelvar's thoughts and actions. I was reading various well researched articles about the emerging future. Also let us look at various schemes being thought of or launched to reach the benefits to the poor e.g. Jan Dhan bank accounts; linking accident insurance with that; plans to link subsidies directly to people; ideas about Digital India. None of these benefits can be accessed only with a thumb impression. Also for a farmer or artisan to get a better price for his/her products, learning more about the markets is vital. The levels of cleanliness that will be insisted upon in various "simple" job markets like working in restaurants or hotels (using a fashionable term "hospitality sector") require much more knowledge and continuous learning.

In about two decades from now those Indians who cannot use computers, smart phones etc., will be washed away as "illiterates" though they can read and write their language. I myself after many decades of experience and learning, am now trying to learn many new computer and ICT based skills, to access new information. The amount of current research material and review documents, I access though the internet is large. They are vital for writing these articles or other books; to give meaningful talks with updated information; many of which are not available in print.

I therefore came to the conclusion, Education in all forms of modern knowledge bases, and related skills are vital for every person living in the modern world. That education should be such as to make each individual a continuous learner all through his/her life, as technologies, business, the jobs, the governance systems, social interactions, resources, the laws etc., are continually changing. **The progress which is based on new knowledge and skills, is also making us obsolete in a short period.**

Within a generation, with active life span extending to 85 years plus (soon to be for most Indians) one will get obsolete about 8 to 10 times in one's life. This includes teachers, doctors, engineers, scientists, artists, accountants, administrators, yes, in all professions, including the "ordinary" levels of economic activity like farming, water management, policing, maintenance, etc.

Is India Prepared?

Is India or Indians prepared for it? I have to say no. "Education for All" is another fashionable statement. "Global Ranking" is a new fad with which academics are obsessed with and they are pushing Indian elite institutions more away from India related problems, in order to get a better citation index in an international journal with higher impact factor (that has become full time goal of many academic leaders, administrators and academics whose promotions depend on these).

Falling into these traps, Government keeps on announcing more of IIT's IIIT's, AIIMS like medical institutions, IISER, NIT's etc. Thinking at the Government level be it Delhi or state capital. It seems to be "you create more Government schools or institutions and/or increase funding for Government schools/

colleges/universities"; then India's educational problems will be taken care of. The reality is far from that is that Government funded/aided schools/colleges/universities including elite IITs' IISERs, IIMs, IISc, AIIMS etc., cater to a very small percentage of the total students.

Also Government statistics generators are happy with entry numbers Gross Enrollment Ratio (GER). Let us have a reality check. Let us look at a research report "Internal Migration for Education and employment among youth in India" by S. Chandrasekhar, and Ajay Sharma (Indira Gandhi Institute of Development Research, Mumbai January 2014 http://www.igidr.ac.in/pdf/publication/wp-2014-004.pdf.

The perceived failure of India's education policy to arrest dropout rates and deliver quality learning along the various stages of education ladder is empirical fact. While India has steadily moved towards universal primary education, the age specific attendance ratios need to be improved. The age specific attendance ratio is calculated by dividing the number of persons in a particular age-group currently attending educational institutions by the estimated population in the age-group 6-10 years and then multiplying the resultant number by 100. In fact, in 2007-08, the age specific attendance ratios were as follows: 6-10 years - 88 percent, 11-13 years - 86 percent, 14-17 years – 64percent, 18-24 years - 18 percent and 25-29 years - 1 percent (Government of India 2010). There are also considerable variations in the age specific attendance ratios across of India.

The least gains have been recorded among those in the age group 18-24 years - from 10 to 15 percent in rural India and from 23 to 27 percent in urban India. In contrast to the age, specific attendance ratios where we do not take into account which class or grade the individual is attending, the net attendance ratio is defined as the ratio of number of persons in the official age-group attending a particular class-group to the total number persons in the age-group. The net attendance ratio drops off sharply after class V and is only 8 percent among those pursuing post higher secondary education (Table-1).

Net Attendance Ratio by Broad Class Group (All India)			
Rural Urban Rural + Urban			
Class Group	Total	Total	Total
I-IV	84	85	84
VI-VIII	57	65	59
IX-X	38	51	41
XI-XII (general Education	22	39	27
XI-XII* (all education)	23	40	27
Post higher secondary (general education	6	14	8
Post higher secondary (all education	8	21	12

** includes diploma with minimum requirements below higher secondary*

Education is categorized in three classes in the survey: (i) general education, (ii) technical and professional education and (iii) vocational education. All education includes (i) (ii) and (iii)

Source: Government of India (2010)

"A scenario where the net attendance ratio at higher levels of education can be doubled would augur well for India's youth and hence for the prospects of the economy. The fact that the East Asian countries managed to achieve a sustained high growth rate beginning the decade of sixties is often attributed to their singular focus on three outcomes, viz. improving educational attainment, increasing workforce participation rate and stepping up the higher investment rate.

While India's has a healthy savings rate of 34 percent and investment rate of 36 percent (Government of India, 2010) it still lags in improving the quality of human capital and increasing the workforce participation rate, and in particular that of women. There is evidence to suggest that educating and skilling India's youth by improving access to tertiary education and increasing the completion rates have significant implications for the economy".

Government Response

It is certainly true that large sections of the population in India do not have access to quality education. Every year, at the time of college admissions in Delhi, there are reports in newspapers about the extremely high 'cut-off marks' set by colleges for students seeking admission. In some streams, 100 per cent marks in the school-leaving examination are required for a student to gain admission to college. We thus have the somewhat ludicrous situation that a student who has got 99.9 per cent marks in the school leaving examination cannot even be considered for admission to a college. It is true that hundreds of students do boast 100 per cent marks; but where are the remaining students to go? The problem is compounded by the fact that the colleges under Delhi University are funded by the University Grants Commission (UGC), and hence the fees students' fees in these colleges are nominal. If a student goes to one of the private colleges outside the ambit of University of Delhi, they have to pay much higher fees and also do not get the coveted University of Delhi degree.

It is not that these problems of exclusion exist in higher education alone. Admission of small children to nursery schools is another area that is fraught with tension. In response to several cases filed by parents alleging arbitrary selection methods, the Delhi High Court had to set a number of criteria for admission.

Meanwhile, the Right to Education (RTE) Act, formulated with good intentions, has also been causing problems to the neighbourhood private schools that cater primarily to children from lower middle class families. The RTE Act has enabled state government officials to intrude into their functioning as inspectors. Several private schools have closed down as a result.

Perhaps a thrust on the betterment of government schools through a vigorous pursuit of the Sarva Siksha Abhiyan, rather than the inspection of private schools, would have served the children of the lower socio-economic classes better.

The government typically reacts to the need for a more extensive education system that can serve many more Indians by levying an educational cess over and above the income tax that citizens pay. But when it comes to expenditure on education, a fairly large fund goes to the creation of new IITs, IIESRs, and NITs etc., often without adequate preparation. The money could be better utilized in equipping schools and polytechnics with skill-imparting institutions. But such an approach is not thought of because imparting skills actually comes not under the Ministry of Human Resource Development (HRD) but under the Ministry of Labour. It appears that the new government is looking to combine the two. But it is not enough to integrate departments. It is crucial to look at skill development at every level of education.

One of the results of the difficulty of getting admission to good colleges in India is the trend on the part of parents to try and send their children to foreign colleges and universities to complete their education. Most middle-class parents, of course, cannot afford the high costs of foreign education. But the aspiration to send children abroad for education, sometimes without tangible benefits, creates unnecessary financial pressures on the parents, a drain in the foreign exchange reserves, and a false social disparity between those who can afford to pay and those who cannot.

Even more worryingly, coaching classes becoming the norm actually indicates a failure of the education system, which is clearly unable to prepare students adequately for professional degrees or competitive examinations.

Crucial Statistics

The seriousness of the problem becomes clear when we look at the numbers. Parents who can afford to pay for their children to go abroad collectively remit about US$ 6 billion (about Rs 36,000 crore) abroad every year. The same parents would have spent about Rs 4,000 crore a year to prepare their children through coaching classes. At least the money spent in coaching classes ploughs back into the national economy and creates jobs. But the money that goes abroad is lost to India, and more often than not so are the students, many of whom do not return to India after acquiring a foreign degree.

We are referring in the above, the students who are qualified for higher education. Much worse is the situation of a large number of students who do not get the chance to get a good education at the primary or elementary stages. About 70 per cent of all students in India stagnate as this stage. In government data they are technically called literate, but they cannot read or write well, nor do they have the requisite social or technical skills to compete.

In the age group of 15-18 years, this student population adds up to about 56 million. About 2 million of these students might have relatively better skills.

In the age group of 22-25, the total number of such students is again about 56 million, a very large population of young people completely unprepared for the challenges of the modern economy. About 2 million of these youths, who are relatively better prepared, may get into the IT and financial sectors, but even many in this 2 million do not get the jobs that they aspire for. Many more are left unemployed.

If we take into account all those who get a college degree, the number we are looking at in the age group of 22-25 is 8 million, students who come out of the 40,000 colleges in India with largely inadequate social, economic and technical skills.

The two tables below (Table 2 and 3) present all the relevant numbers for these two age groups of students. (Figures estimated by the author from Government. data and other public information.)

Table 2: Human Resources in India (Age Group 22-25) @ 2014

a.	All India Total	80 million
b.	Those with reasonable skills and higher education	2 million
c.	Those poorly equipped yet having a degree	6 million
d.	Those without a degree, may have passed school, poorly skilled	16 million
e.	Those who may have some primary/elementary school education, poorly skilled	56 million

Table 3: Human Resource in India (Age Group 15-18) @ 2014

a.	All India Total	80 million
b	Those who have done well in 10+2 and hope for better studies (professional, degree, etc.) with reasonable skills	2 million
c.	Those who have done 10+2 schooling but can only aspire to get into one of the 40,000 colleges, which do not prepare them for higher modern skills	6 million
d.	Those who do not hope to get studies higher than the 10th class	16 million
e.	Those who have dropped out before the 8th class	56 million

For those in categories (d) and (e) in Table 3, the situation is particularly bad. Over the past two decades, many of these persons have gone into life unprepared. Those who were in the age group of 15-18 in 1991, the year of the liberalization of the economy, will be in the age group of 38-41 in 2014 and struggling to meet the basic needs for themselves and their families. They will typically be in the unorganized sector, often with an irregular job, leading a day-to-day existence. They would aspire for a better future for their children, but how can they build that future?

In fact, the fate of the children of those, who might be in the age group 15-18 in 2014, is almost shockingly similar to theirs: many of these children would fall into the categories (d) and (e) in Table 3. These are, however, post-liberalization children, exposed to the global consumer culture, harbouring very high aspirations. They want to have a life similar to that of their cohorts who sit in category (c), or even (b), in Table 3. **But is there a chance for them to improve their life even in a decade's time, by 2025, when they are still under thirty years of age?**

So far, the government has only been able to come up with stop-gap solutions to address this vital question, in the form of food security for all or a minimum guaranteed employment of hundred days for every person. These ambitious schemes are difficult to implement, and even if they are implemented, they do not make for a productive action plan that will help develop and sustain the national economy. Some states give freebies.

The real solution is to ensure that all those in table 3 in categories (c), (d) and (e) acquire the high skills required for the modern economy and earn more through the productive activities unleashed by rapid economic growth. In fact unless these people are relevantly skilled very fast, India's economic growth will stall. **Action needed is immediate!**

That is the challenge before the Indian education system. What should the action plan for education be?

Does the solution lie in stepping up the government budget for education? Not exactly merely pumping funds into building more schools is money-intensive and might not be very productive. It is reported that in order to comply with a provision under the RTE Act that no child should have to travel beyond 1 km to go to school, a number of government schools have been constructed and the number of students admitted in these is very low. This is a terrible wastage of taxpayers' money and does not benefit anyone.

Here are our recommendations to reboot India's education system.

For Schools in Villages

There is a simple scheme with far-reaching impact called Providing Urban Facilities in Rural Areas (PURA) that has been propagated by Dr. Kalam for some time. Its model for improving the rural education system is clear and precise. Working with a network of about fifty villages that are fairly close to each other, the aim is provide physical connectivity, electronic connectivity and economic connectivity to this network.

Where physical connectivity is concerned, **if the fifty villages are connected by a reasonable road, all of them together need to have only one high-quality school with a number of good teachers teaching a variety of subjects, rather than several smaller schools in individual villages with limited teacher and resource strengths.** The Pradhan Mantri Gram Sadak Yojana (Prime Minister's Rural Road Scheme), started in the early 2000s, can help in ensuring the road connectivity.

This can ensure the reach of tele-education to the children, using satellite and broadband services to connect with resources, data and instructors around the world. Multimedia and interactive channels can impart relevant skills to children right from the primary school stage.

All this connectivity will, however, be meaningless if the curricula, syllabi, pedagogical methods, teachers' knowledge, etc. remain as they are now. Most schools in India use a unified curriculum. This one-size-fits-all system of education excludes many children from the benefits of education. What is needed is to provide many more options for schools and teachers to select from.

The primary and elementary stages of school should be the time when children are given an orientation towards the hundreds of skills that are required and relevant. They should be made aware of forms of learning beyond the three R's: reading, (w)riting and (a) rithmetic. With such an exposure, when they go past elementary school (eighth standard) and reach fifteen years of age, they will be confident enough to choose their preferred career paths, from multiple options.

To enable this transformation, what is needed is also a teaching and training mission. We need to find innovative ways to create new teachers, going beyond our dependence on teachers with degrees in education training. A large number of educated women drop out of the employment stream after childbirth because they cannot adhere to extended work hours. Many of these women, who are based in small towns, can be deployed for part-time teaching. The same is true of our senior citizens, who are now 100 million strong; many of them are skilled and well educated. Instead of emphasizing formal qualifications, we can also utilize the skilled workforce in the country for teaching. All of these teachers can be accessed by the tele-education mode from villages.

For Schools in Towns and Cities

What we have said for villages can be adapted very easily for towns and cities. Again the emphasis should be to collect networked students into a

good school rather than spreading them out. When looking at the urban part of the education system, the government must be careful not to discriminate in favour of government schools and against private schools; it must assist all schools with materials and facilities equally.

In urban and semi-urban areas, **colleges and nearby schools can be converted into powerful knowledge-skills-imparting centres, and engage in distance learning through broadband connectivity.** The broadband connectivity should aim to transform our 40,000 colleges at the first instance from institutions steeped in the conventional rote learning mode to modern global-level skill imparting centres. Again a transformational change in curricula, syllabi and pedagogical methods will be essential.

A Time for Change

What we have discussed so far applies primarily to the children in categories (c), (d) and (e) of Table 3 aiming to transform them from their current state and thereby create a formidable human resource for India.

When the levels of children in categories (c), (d) and (e) of Table 3 are raised higher – and this can happen in a few years, by the year **2020 – Table 3 in say 2025, a decade from now, will look very different from what it looks like today. All of the people in categories (c), (d) and (e) will be highly skilled. Category (c) and (d) will each have about 30 million people, with category (e)close to zero. Meanwhile, the numbers in category (b) will have risen from the current 2 million to about 20 million. That is how the supply chain works.** If we attempt to raise the numbers in (b) without improving (e), (d) and (c), how can we have the excellent inputs we will need to support the larger numbers at the higher level?

That, in a nutshell, is the problem facing professional education today. Opening up more and more IITs and IIMs does not solve the problem; it only dilutes the brand equity of these institutions. As we have said, the focus needs to shift to providing the relevant skills and knowledge to students at the primary, secondary and college levels. Alongside, there is a need to create many more educational institutions in India, so that more options are available to students. For this to happen, **India should remove all the fetters against foreign direct investment in education. The government, on its part, should stop subsidizing the cost of education and instead provide widespread scholarships to the poor and needy, so that they can benefit from the best that modern education systems have to offer.**

This change of mindset would not only help the many millions of students who are currently stuck in categories (c), (d) and (e) in Table 3, but also liberate the 'high flyers' in the elite institutions who are currently trapped in their own private rut of excessive expectations and limited scope, which stops them from performing to their fullest potential.

In the right environment, they will flourish, as the other parts of the student spectrum blossoms. Original creative work, new discoveries and above all new inventions and innovations will then emerge, with India-invented, India-designed, India-produced products and services spreading all over the world.

The day this can happen is not far away. A decade from the year in which we liberate our education system from government control and enable the inculcation of relevant skills and knowledge at all levels, India will truly excel.

Let the starting year be 2017, so that the year 2027 brings glory to India.

What should be done?

For Village Schools:

The most important action to be taken to provide all villages remote areas with:

- Good physical connectivity (roads or rails or even waterways)
- Good electronic connectivity (to jump the many lost opportunities a 3G type of connectivity)

For those of Indians who get into much higher studies and research, in addition to freeing the existing elite institutions under government, we need to encourage them to fly high. It is described in detail in the book Way Beyond The Three R's: India's Educational Challenge in the 21st century" by Y.S. Rajan (2010, Penguin). It elaborates how to cater to these avantgarde persons and how to enable world class institutions to emerge. This will be in all disciplines of knowledge including arts, literature, philosophy, law, archaeology, history etc., in addition to science, technology engineering, mathematics, and medical sciences.

❐

17
Migration of People

Migration from the place of birth has been important feature of human evolution. It had its rewards and pains. After the invention of agriculture around 15000 years ago, people started settling down, especially around river beds and other areas having access to water. Forests were another preferred habitat. Most people started creating villages and towns. Thus began great periods of history: kings, emperors, kingdoms, and great works of arts, literature and philosophy. Later various discoveries of science and technology emerged: astronomy; fire, medicine, metallurgy, bricks, other building materials, use of animal power, carts among many innovations.

Even then, there were attacks from other kingdoms, from nomads outside the kingdoms etc. Many people had to leave the places where they settled. Also vagaries of weather and pestilence caused migrations of large number of people.

During the periods when Europeans colonized many parts of the world, they took slave labour from one place to another, through ships, railways etc. Such immigrations which had taken place during the past few centuries due to colonization, slave labour, religious prosecution etc., have left many "historical wounds" often unhealed. They erupt not infrequently in different parts of the world: colour prejudices in USA, many conflicts in Africa, Middle-East, Asia and even in Europe. In India we have many pockets of such troubles. Recent events in France are yet another example.

Migration within India

In addition, in the post-independence India, economic growth and regional imbalances as well as the "historical" problems and the poverty of social and economically deprived castes and tribes, led to migration within India. Life of such internal migrants is not easy. Many of them are exploited inspite of the plethora of laws "giving constitutional protection and equality". There are many conflicts within these groups as well; they do not integrate. Their divisions are exploited by the more privileged.

In addition, there are many illegal immigrants from the neighbouring countries. Thus there are often both ethnic and religious tensions. But these illegal immigrants as well as the Indian migrants (intra-State or inter-State) play important economic role. They often do jobs which the local population is unwilling to do. In States which have become prosperous through agriculture/industrial routes often attract immigrants, due to economic opportunities and lack of local labour. Green revolution areas attract the most. The case of Indian immigrants to the Gulf countries and Middle – East since 1970's, is a classic example. It has benefitted those countries and also India. Though some of the immigrants suffer in such immigration, by and large many of them have uplifted themselves socially and economically. But it is not always the case with the immigrants within India. Though they earn more than what they do by being within their villages/towns, improvement in their economic status is slow. The worst sufferers are the children who miss their education. However they are able to be in touch with their villages due to improvements in road transport (now 85% of passenger traffic); railways are only 14% as against 85% during 1950-51. Still most

Government schemes do not benefit the migrants, as they fall between many "stools" of administrative identity cards.

Migration and Human Development in India

We draw much of data and information from the United Nations Development Programme (UNDP), Human Development Reports Research Paper 2009/13, April 2009 by Priya Deshingkar and Shaheen Akter, with the report title as in the title of this section.

As per the report, the "migrant labour makes enormous contributions to the Indian economy through major sectors such as construction, textiles, small industries, brick-making, stone quarries, mines, fish and prawn processing and hospitality of services."The report states "But migrants remain on the periphery of society, with few citizen rights and political voice in shaping decisions that impact their lives". It goes on further to inform us that "unlike countries in Southeast Asia and East Asia, the bulk of the migrant work force in India has little or no education. In fact the migrants are poorly endowed all round".

According to the Census data for 2001, 30% of the population or 307 million are migrants. The UNDP report points out these are not all employment related. About a third of them had migrated about a decade ago. Change of birth place and change in the last usual place of residence is used as "migration" and hence movement of women after marriage also gets into such statistics. Thus policy making becomes difficult as the report points out other short comings of such official data. Among them are the "Circular migrations" rural to rural and rural to urban which are short term migration for work in farms and commercial work and return to villages. But such migrants lose out of many benefits of government schemes and especially education of children. Healthcare access is another major issue for them.

Some of the detailed estimates of the report give the following information (we have collated them)

- Construction section	:	At least 30 million
- Brick kilns	:	At least 10 million
- Textile Industry (largest Export earner	:	About 35 million
- Leather	:	About 2 million
- Other small industry/ informal sector	:	Not given as there are millions of such informal sectors.
- Mining	:	Around 3 million (artisanal and small mines)
- Quarrying	:	About 4 to 5 million in Maharashtra alone
- Mica Industry	:	Not given as it mentions gross under reporting due to labour laws.

Agriculture, Sugar Cane, Cotton seed, Cashew processing, restaurants, domestic help etc., are given but not covering all. We have not given all items referred to in the report.

The report states that "adding up the numbers above in the sectors reviewed we estimate that there are around 100 million circular migrants in India today". They play a huge role in Indian economy but that is not recognized due to lack of data (often obfuscated due to complexity of labour laws) As can be seen in a data given later, there is annual growth of about 12% in informal workers and this number of migrants may now be around 150 million.

Numbers and GDP Contribution

The fate of such migrants is closely linked to the informal workers, who lose out on many benefits. During 1999-00 as per National Sample Survey (NSS), their number was 361.7 million (given in the UNDP report). During 2004-05 it rose to 422 million, which is approximate an annual increase of 12%. Thus through extrapolation of the same trend we can derive (we have derived it) from the Table in the report given upto 2004-05) for 2014-15.

Informal workers working in informal sector	–	521.73 million
Informal workers working in formal sectors	–	58.62 million
Total informal workers	–	**580.35 million**
Formal workers working in informal sectors	–	1.4 million
Formal workers working in formal sector	–	32.8 million
Total formal workers	–	**34.2 million**

The report states: "Overall data show that there are more informal workers in the rural areas than urban and among the rural workers more women than men". This shows the vulnerability within this group; migrants amongst these are much more vulnerable. Migrants amongst the informal workers would be about 25% (see the earlier section).

After going through a number of tables with disaggregated data for various sectors like agriculture, mining and quarrying, manufacturing etc., the report concludes: "The contribution of migrant workers to the Indian economy is at least 10% of the total Indian GDP" It is a conservative estimate as the authors did not have data/estimates of all sectors.

During the economic slow-down, the migrant laborers are the first to lose their jobs. They do not have any protective cover.

Human Development Impact on Migrants

In contrast to the workers in micro and small industries who do not have any job protection and work like casual workers, migrant workers, earn less than the regular workers of micro/ small industries.

While there is a marked discrimination against migrants in terms of payments, still there are some case studies which indicate that about 10% in Bihar, use the circular migration route (to jobs in Mumbai), lease some lands back home with the small surpluses they create from their meager incomes. Thus they do some additional farming through other family members and earn a bit more than they would, had they stayed in villages.

310 families from rural UP, in another case study, indicate a shift towards better life. 84% women interviewed indicate better access to education for their children.

There are also negative side reported that "migration carries risks of injury, exposure to life threatening diseases, loneliness and increased work burdens for women left behind".

There are a few other social dimensions:"the scheduled castes and tribes said that working outside the village had given them the opportunity to work with dignity and freedom. Bihari migrants said that caste hierarchies in rural society were rigid and humiliating".

Benefits from Government Services

Almost all migrants are excluded from Government services due to non-possession "valid" identity cards: ration cards, BPL card to avail of medical benefits etc., though they are often the poorest of the poor. Their children cannot get the benefits of mid-day meal scheme in Government schools, even when some try to send their children.

Also they do not have any banking services and have to depend on the middle men to give them advance money before he takes them to the contract work. The payment of interest on loans is an additional burden. When we started reading the detailed case studies, it is found that these middle men at least help better than the government systems like local Panchayats, as often caste factors and related humiliation work harshly in such systems.

A study in six districts of Maharashtra (2003), have used the Sarva Siksha Abhiyan (SSA) implemented through the Mahatma Phule Sikshan Hami Yojana. They had covered 200,000 children of the migrants. There may be such schemes elsewhere.

These schemes should be emulated and if necessary augmented by satellite TV, as education of the children of the migrants is crucial for the future of India. Some of these children may assist their parents in their work. Let us not be dogmatic and invoke child's rights and stop them. They get skilled in the process. Some work experience in addition to education is good for these children.

None of Food security schemes helps these migrant labour. In addition to the fact they get lesser incomes than the non-migrant informal workers, they have to buy their essential food items in open markets, in small retail shops. Thus they lose a lot.

Of all Indians these 150 million odd Indian migrants (circular or otherwise) who require UNIVERSAL SIMPLE IDENTITY cards for themselves and their children. Let us not worry too much about the abuse of cards and make the system complex. Let us also be sensitive to the social cruelties in the villages and **not vest too much power with the Panchayats in the issue of such cards,** as many of these migrant workers are from low scheduled castes and tribes.

These ID cards should be useable in government hospitals, ration shops, schools and other schemes benefiting the poor. **Can PM's Jan Dhan Yojana be applied and they are given a Ru Pay card in which the contractor who employs them deposits advance money and also the wages later.**

Also can the digital India programme create a website to offer the services of such migrant workers? If some IT company operates such a facility, their expenditure may be treated as CSR.

Why Worry about Migrants?

Migration and human mobility are not an isolated Indian problem. There is a UN system Task Team on the POST-2015 UN Development agenda on this issue. A report on this Agenda notes that one in seven people live outside their country or region of origin. Again many of these estimates do not include many persons migrating on a seasonal or temporary basis.

Total Population of the more developed regions is unlikely to increase: from the current level 1.2 billion they may reach 1.3 billion in 2050. They will experience shrinking workforce as many more elderly persons will live longer. Therefore various forms of international migration would take place-both legal and illegal.

From India, better skilled Indians would like to migrate to countries with better economic opportunities. They face one-type of problem which will occupy the attention of government. It is already happening in India: the fate of entrapped nurses in Iraq or Syria or the problems of Indian workers in the Gulf countries or "racial" attack cases in some developed countries. They have some voice: in media, concerned Governments in the states from which they have gone etc.

The UN Task Team would set up some agenda after several years of discussions as to how to deal with the international immigrants – their rights etc. But it is unlikely that they may take up internal migrants except in terribly war torn countries in which they are more like refugees.

But the internal migrants for a longer term or through seasonal/circular migration routes will continue to increase in India. Development of regions in India is uneven. This is due to various historical reasons, local politics and present governance issues. States, which are poorly governed, and have poor infrastructure, unskilled human resources, law and order issues etc., are unlikely to attract private investment. Also even the public funds are unlikely to be utilized well, as we have witnessed over the past six decades after independence.

With global and domestic market forces at work, some states may reach much better standards of living and have much better economic opportunities as the need and the potential to grow more will be there in India till 2050. These States will attract more migrants from other States of India and even from other developed countries. One category will be those with good quality higher education and those with high specialized skills be it engineering, medical fields, security, accounting, foreign languages etc. It will be easier for persons from this category to decide to stay in these states; even for foreigners. (It is already happening in many metro cities of India). Such persons are also globally sought after and they may migrate abroad. Some may do new forms of "circular migration"; work in developed country which pays much more and come back to India for 3 or 4 months. One can do tele-work with various ICT means (including video conferencing) and also enjoy one's culture, friends, relatives etc. It is unlikely that these persons face many problems themselves. Their better life-styles may become a cause of "jealousy" in some. But by and large such populations which may be around 50 million by 2030 will be one set of migrants.

The other category will be a large number of poorly skilled and far less materially endowed persons from the less developed States of India. They will fill the huge demand for "low paid" jobs which would be shunned by the "locals" of the State: garbage collection, municipal services, domestic helps, petty security services like being a watch man, etc., as well as many informal work in the micro and small enterprises. It is unlikely that "casual work" will disappear from the Indian economy as it has deeper economic benefits for the privileged classes. Number of such persons will continue to be a few hundred millions. They would mostly do "circular migration" – going back to their own poorer states for a few months in a year and return. No doubt there will be incremental improvement for them in their "own" villages and "towns" as they will remit some surpluses from their earning in the outside States, back "home".

The negative aspects are: most of them will feel rootless. Their own States are floundering through slogans; and partisan politics may be widespread. They have seen the realities of what is possible, in other States. So these slogans cannot be exciting to them, as to those who are left behind. Their neighbours who did not migrate would feel jealous of them. Those who moved from lower classes would feel that the "village politics" does not give them "higher status" despite their better economic standards and skills. "Democratic" politics of India has strengthened the caste inequalities instead of removing caste feelings. In the developed states where they go and work, they are in the bottom of the pile. Though not humiliated as they are done back "home" in their villages, the anonymity of city life in the developed State leaves them with better standards of living but without dignity.

The migrant situation ensures that their children would be condemned to the same life if our policies remain the same. (Please recall our narrative on lack of educational opportunities for migrant children).

Above all, with none realising (that is the elites of the developed States, middle classes and the migrants themselves) the migrants with poor conditions of living would also be carriers of many deadly microbes. Microbes don't distinguish between human bodies (rich or poor). Rich and upper middle classes (UMC) might have "protected" themselves against many known bacteria and viruses (microbes). Some other microbes in the bodies of these poor may find their way to the bodies of the rich and UMC. Nature does not differentiate the human-made classes.

Rich and UMC's cannot be locked into their palaces and villas. They would like to enjoy the outer world, including many habitats having the beauties of Nature: forest resorts, boating, trekking etc. So the health of every citizen is crucial for the health of all. Access to health care for the migrants becomes crucial agenda for all including Rich, UMC's and middle class.

Another factor is social feeling. No society can afford to have fairly large sizes of population feeling rootless. That will be the breeding ground of petty crimes to extremism. This is yet another important reason why India (for that matter any country) has to worry about the education and socially uplifting activities of the migrant population.

Living with Nature in the modern world, provides many beautiful bounties for the human beings. The levels of happiness will be much more that what it is now, where most people have to be locked up in huge apartments, face congested traffic, and polluted surroundings.

Above all, the disconnect between our evolutionary inheritance (love of green in the forests, the lively humidity of village life, the cool nights of deserts, the coolness of water bodies, the music of flowing water, the vastness of the oceans, the imperious heights of the mountains etc.) and the artificial modern life is a serious problem for all. A few Rich can try to bridge it; but it is incomplete only when all members of the society are able to participate in it. Then alone the evolutionary inherited emotion is fulfilled.

Conclusion

For the first time in human society, it is possible to achieve this for all human beings. But we need to carefully to look at all segments of our society (true for all countries including India) Rich, UMC's Middle class, minorities of various forms, as well as MIGRANTS of various forms, described above. Most vulnerable among the migrants are in the lower range of the pile. Let us care for all of them in the realistic form.

All of us can enjoy Nature in our modern world.

❒

18
E-Commerce and Virtual World

The Indian consumer, especially in the metro cities, seems to prefer purchasing through e-commerce routes. A quiet (and also not so quiet due to an incidence with a company) revolution is taking place in terms of booking for taxis. I had talked to a number of consumers of such services in Bengaluru and we have used it too. Ola has "aggregated" taxi services and a person can get a taxi within any place in Bengaluru city within about 20 minutes by calling or using an Apps to book Ola. The taxi is used mainly for a pick-up and drop. Asking to wait and stay is costly as time clock ticks fast at the rate of a rupee! But pick-up and drop, is very competitive compared even to auto rickshaw fares, so what is the problem if waiting is costly? One can finish the work at leisure or hurriedly and book for another taxi at another place where one is.

What is the impact on the drivers? They are all very happy. They don't have to overwork all day. They can fix the time of their availability and be available at that period. No hassle of bargaining; the amount comes in their mobile. Take the amount and go for another duty. Many drivers told me that they get about Rs. 3000 per day net profit. There are also other drivers who do the earlier conventional duty of pick up in the morning, wait in office, and drop in evening. Then they do "Ola" duty for a few hours at night and/or weekends and earn more.

Passengers are happy; the drivers are happy. The middle persons who used to run an office, receive calls, fix rates etc., are now the most affected, as drivers directly book with Ola, complete the formalities of police check etc., get the standard instruments for e-contact. Many of the small scale businesses of these "middle persons" who run taxi service businesses,have their own cars or have drivers with cars registered with them, have to charge their overheads on the consumers (passengers) and on the drivers. Those overheads are now in e-aggregation mode spread over several thousands of drivers, unlike the traditional taxi service providers who used the traditional mobile phones and did business by talking. They at best texted (SMSed) driver name etc. They had around 50 to 100 taxis at their command some even less number. The new service mode of e-commerce (Ola, Uber etc.) "aggregates" several thousands of drivers and has complex computer models which link the nearest driver to the passenger seeking the service of a driver (car sizes also can be chosen).

Passengers link to this "aggregated information" available in the virtual world and get the services directly without direct intervention by a "physical middle person". The virtual world of computers, networks software and mobiles, do their linking. Passengers can choose the car type, in whatever place they are going to stand. **This method of e-commerce using the virtual world for selecting the goods or services through Virtual Medium has the economy of scales.** The managerial overheads are spread over many thousands of people. Therefore the services are competitively priced for the consumer and the actual person who delivers the service is also paid well. "WIN-WIN" for both these group of persons. **But the traditional service providers are losing out on their businesses.**

Remember "good old days" as late as 1980's when telephone operators with excellent voice and diction were sought after by the telephone exchanges and offices and were paid fairly well. With the advent of automated long distance trunk dialing, their importance started waning. Nowadays most big organizations have just one operator. Most of the enquiries are done by pressing one number or another, and operator is the last resort help desk.

That is the power of technology. Automated exchanges use computers to route calls. Later came train booking. Now(2015) reservations for railway tickets including selection of seats are done through internet and also paid electronically. It is also an e-commerce in the virtual world.

We explore in this brief article, the dimensions of e-commerce as they are now emerging in India and the new possible directions. We will also look at the advantages that can directly accrue to rural economy and those who depend on agricultural produce for their incomes.

Why E-Commerce is Getting Popular?

For airlines booking through internet or smart phones and for tele-check in, the advantages are very obvious. Earlier, ticketing facilities used to be available in a few places in a city.

So is the reason for rail ticket booking and also for many tourist bus services – for a city tour, inter-city tours, tours to places of religious pilgrimage etc. There are many one-person businesses as well which help a number of not-so-tech-savvy passengers including those in lower middle class who may not have internet or smart phone facilities, to book trains, tourist buses with a small service change.

For these travel related services which sometimes include hotel bookings as well (for tourists), the advantages are very obvious. No hassle of long queues. This also helps them from the loss of leave or wages, which they lose, if they have to go to railways station or select offices to go and get tickets etc.

Nation benefits as a whole, because the economically productive persons, do not waste their working hours inefficiently by standing in queues. For getting such efficiencies for the national economy a percentage of persons in airlines, railways, and bus services who were doing a "parasitical" task of issuing tickets one-by-one (a large inefficiency) would have "lost" jobs. But as the overall efficiency of the system improves with many passengers who travel, these agencies also gain a lot. The redundant persons who were performing inefficient and parasitical jobs would be deployed to more productive tasks.

In a similar way, look at the advantages of ATM Services provided banks. The economic efficiency, these ATM services have provided for ordinary Indians and the economy is many-fold. In addition, the well-being of people is increased several-fold, as they don't have to wait for a bank to open for cashing cheques and almost losing half-a-day in casual leave.

Internet and mobile banking have increased many times more the conveniences of banking. Of course, this is still limited to a small percentage of Indians, who are educated and take advantage of internet penetration; there is some good news that rural internet penetration is increasing.

Credit cards given by the banks cover still more affluent sections of Indians. But this service has added a great convenience for an Indian to shop, to go to hotel or to book a ticket etc., Under the Jan Dhan scheme and the RuPay credit cards under the scheme, the current Government of India, intends to extend the convenience (albeit in a limited way) to all Indians including those in the Bulky Base.

Another group of middle class and elite group Indians who benefit from such e-transactions (virtual world transactions) are those who pay Income Tax, Service Tax, Property tax etc. Though, there are a number of problems of transition from the earlier system of individual paper filing (more so because of the old mind-set of the tax officials), this e-system is of great help to tax payers. Here again the benefits are obvious: One does not have to run-around and waste time on enquiries.

We have described in some detail in the earlier section about the advantages of booking taxi through aggregated systems like Ola, Uber etc.

We have not included in this set of e-commerce/Virtual world, the traditional forms of phoning and booking taxies, hotels, and even ordering of food. These are mainly due to easy availability of telephone connections (land or mobile) food delivery is simplified because of easy transport (two wheelers) and food packaging facilities (from plastic bags to heat conserving packages). Of course, these add a bit to the overall price charged to the consumer (but the convenience is severalfold).

Beyond these items, e-commerce is mainly around purchase of goods and services through ordering in the virtual world. These are also called e-retailing or e-tailing.

E-Tailing

In a recent article Saturday 21, March 2015, in Business standard under the heading "Flipkart has biggest piece of Indian e-tailing pie" there is a brief report of the new report by Morgan Stanley.

The market share of companies is as under:

Flipkart	–	44%
Snapdeal	–	32%
Amazon	–	15%
Others	–	9%

In terms of products/services sold in Indian E-commerce the article states:

Electronics & appliances	–	51%
Fashion	–	30%
Books	–	7%
Others	–	12%

There are a number of Merger and Acquisitions (M&A) of different companies which are creating new business opportunities. For example, most of Flipkarts' business had been from consumer electronics, books and other categories. With the acquisition of Myntra which has advantages in the Fashion Space, Flipkart became a market-leader.

The coming months and years would see many such M&A. Such M&A are taking place in the taxi service providers as well.

This business segment is emerging as a big business with veterans like Ratan Tata investing on many such startup companies.

Is it only Upper Middle Class and Elitist Zone?

Let us look at an article in Business Line 23 March 2015, under the title "Wealth at the Base of the Pyramid". The subtitle of the article states "World Bank data showcases opportunities in this$5 trillion consumer segment."This is of course an estimate of the world market. In India too it is sizeable if systems are created to create the right type of Virtual Market for different market segments and also create the supply chain for reaching the consumer speedily with quality products.

The World Bank as per the article use **four levels of consumption** to segment the market in each country: Lowest, Low, Middle and Higher, based on income distribution data. The population segments are identified through income per capita. It appears to be similar to what we had described Indian demography, in the Indian context. Rich and super Rich, *RSR), Upper Middle Class (UMC), middle and Lower Middle Class (MLMC) and the Bulky Base (BB) in India would be around 700 million.

As is evident and as per the report, in the lower consumption segments, food and beverages dominate. In this segment, taken together, food and beverages are significantly larger than the middle and higher segments. Higher segments spend much more on services. No wonder the Indian e-commerce business is dominated by electronics goods and fashion as we have given in an earlier section. Finance services and transport also dominate their lives. This indicates that RSR, UMC, and some of the middle class (MC) participate in e-commerce heavily. They have access to internet and smart phone. Their life style demands both partners working full day, suffering all the terrible traffic congestion. They get only the weekends off. They used to rush to the Malls. Again severe traffic congestion is an inhibiting factor. No wonder the e-purchase has come out as convenient mode for buying domestic white goods, fashion dress and even ordering desired food.

The estimate in US $Billion in the Business Line article for the market size of South Asia is:

Food & Beverages	:	646
Housing	:	143
ICT	:	37
Education	:	40

While it is lot better than in post-liberalised India, in terms of quality products reaching rural areas where most of BB persons live and also the Tier II, Tier III towns where Lower Middle Class persons live (to be sure they live in Metro cities too), the benefits of e-commerce have not reached them to a large extent.

It is not just enough to build up supply chains, which are still poor because of poor quality roads, too many tax/check points, poor goods train services. It is also necessary to create the soft infrastructure required for payment for goods (for the seller and buyer).

An interesting recent article in the Business Standard in the Strategist, titled "Playing Catch-up". The article describes in detail how digital payment service companies like PayPal, the internet giants like Amazon Payments and Google wallet and many other mobile payment companies are operating.

The report states that in India "in the mobile payment space, companies have mushroomed all over India. There are about a dozen mobile wallet firms, including Paytm, with 20 million active users (more than the cumulative number of credit cards in India) and others like Mobikwik, Oxigen, Zipcash and Citrus which allows users to digitally store cash (under a cap of Rs. 10,000) typically for low value mobile online transactions". These are good news. Some further facts: "64 per cent of Indians are willing to make purchases on smart phones this year (i.e. 2015). SMS **based payments have the ability to reach out the unbanked sector. Over 40 per cent of digital wallet users hail from Tier II and Tier III towns.** What is more, competitive lines are blurring with telecom companies entering into this arena, an example being Vodafone's M-pesa **Mobile wallet designed to reach out to the under banked or unbanked population.** Thus there is a great opportunity to reach out more of these persons and also to rural areas for reaching agricultural farmers, big and small, agricultural labourers and the migrant workers who mostly hail from villages.

The main spirit of **financial inclusion** desired by the Prime Minister through the Jan Dhan programme can be achieved, if such e-facilities are fully utilized. **The banks need to be creative and reject the notion of "not-invented-here" NIH syndrome and use many of these successful examples.** Government also should not place hurdles on such innovations. Remember how Indian reached TV with 100+ channels to many millions Indian homes through Cable TV (something market pundits around the world did not envision or forecast). Similarly mobile phones themselves can reach such services to all Indians especially in the rural areas. **The parasitical middle persons** (including government inspectors, FCI purchaser etc.) can disappear if the agriculturist is directly linked to consumer via aggregators of e-commerce and the service providers.

Virtual World, Agriculture and Rural Economy

The farmers, not just marginal but also those who are middle level famers, suffer because of vagaries of weather and market. Their options for inputs are also limited. If we list the tradable commodities (illustrative):

- Inputs like mix of fertilizers, micronutrients, pesticides, agricultural implements including pump sets, drip irrigation components and project consultancies, animal/bird health care services etc.
- Outputs from grains to cash crops, horticultural products (where wastage is very high), agricultural wastes, floricultural products, animal, bird, fishery products etc.
- Labour time and period of availability for other sectors of economy, per day or per month wages expected, advance money needed etc.
- Many related artisanal, rural products, herbs etc.
- Tourist attractions in villages etc.

Creative young entrepreneurs may list more of these. In a focused way if these can be brought into the virtual world for people offering to sell and buy, a vibrant e-commerce of rural economy can flow.

I am not unaware of the inadequacies of many elements of rural, town infrastructure: roads, transport, storage facilities, packaging of products with minimum assured quality levels, etc. Packaging is crucial to e-commerce in the delivery (at input or output side). But each of these elements may become major enterprises in themselves in a local area and then expand. See the quality vegetable/fruit outlets in many towns/cities like Pazhamuthir Solai, Reliance, Conduit Fresh etc. There are hundreds of them. How lakhs of eggs are transported daily all over India. Similar aggregations can take place, once a hope and infrastructure are created to exhibit products to the potential markets and for the consumer to order goods/services and pay for the products; the payment should reach the actual supplier almost instantly. Amul's success is based on this. Similar enterprises will develop, for the lower end consumers to niche upper end consumers (organic foods, special herbs etc.)

I hope the persons in charge of Digital India, really capture these real life issues to benefit Indian people. The National Knowledge Network (NKN) is practically languishing for about a decade with very little of optical fibre Network being laid. **Instead of chasing it (it may take years), can India jump start by using a satellite to give such a SATELLITE TO MOBILE** communication service reaching 600,000 villages including tribal areas. Such satellites with multiple beam antennas have just started reliable operations in many parts of the world. If necessary let India purchase one or two satellites from foreign companies and usher rural India into the world of e-commerce. Such a link can incidentally help the mission of universal education (even to higher education) and to reach SKILLS to all Indians. Of course, it depends how well the attention is given to the contents and the language of delivery.

This is the time to "leap frog" the era of traditional super-markets, optical fibres and straight away modernize rural economy through the tools of VIRTUAL WORLD. We will be amazed by the adapting capability and innovative skills of "ordinary" Indian people. For it to happen, the systems of governance should match the speed of new technologies!

❐

19
Softer Issues

We have dealt with many crucial issues relating to agriculture, animals, soil biology, mining, industry, infrastructure, services of various sorts, technologies, human habitats, economics, health, education, skills, governance, possible future scenarios of geopolitics, geo commerce, business management, natural disasters, etc. While addressing these, we did not consider human beings as mere mechanical participants nor as a uniformly automated robots. We did consider the immensity of diversities not only in the biosphere but also in the way human beings live, see things, react to life, aspire for themselves and others etc.

But still, we had not addressed the softer issues about the human organisations, human aspirations, human behavior, culture in detail. One of the reasons is because these are very complex, not yet fully understood scientifically and therefore many opinions are still abound. Even confining to India, the softer issues are not easily describable even for a small State or a district of a larger state. If we add on to the human beings several differentiating layers of gender, class, caste, language groups, property measures of net worth, land ownership etc., and a newer layers such as higher education, education in elite institutions, etc., we will note that the human groups function very differently depending on the combination of or exclusion of many of these features. For example, when super rich groups get together for their 'common cause' they would exclude all others. Even the places they stay are very exclusive. Similarly in geographical areas (even a smaller area) where one religious group has a large majority, other minority groups tend to live in close knit areas, as they perceive that they get a better identity and safety; in some places they are forced to do so due to prejudices. Softer issues for many such differentiated groups are very different: often time in opposition as well.

Some Often Used Terms

The idea that all human beings are equal and have equal rights, gives rise to the concept of DEMOCRACY: "The real POWER lies with the people". They decide how they are to be governed. Such a concept is far removed from the concept of **rule by a king, emperor and their dynastic inheritors.** People are involved in some form or another but mostly they are the subjects of the king. Earlier revolutionary ideas of "people's democracy" also included armed rebellion, civil war, guerilla war etc., to throw away the rule of the dynasties or autocrats and install a rule in the name of people. During the 19th Century and early 20th century, such democracies emerged as an alternate to electoral democracy (that is periodic elections of leaders through secret ballot by people, and COMPETITION amongst various politicians through organised political parties or as independents; this was also called multi-party democracy). The revolutionary people's democracies mostly curbed multiparty politics. The victorious armed revolutionaries established their own system of one-party rule and a form of "democracy" through the party, to "represent" people's interest. Most of these republics have collapsed or have become dynastic dictatorships. Peoples Republic of China (PRC) is an exception. Iran has its own electoral system as Islamic Republic of Iran unlike the western models, the kingdoms in the Middle East and some republican dictators are still left out (which have also fallen into dynastic rules).

India has adopted a DEMOCRACY based on a multi-party system with separation of powers of the legislative, executive and judiciary. There are many checks and balances on the powers of these wings of governance. There is freedom of speech; freedom of protest even by a small minority against the decisions of or proposals of duly elected government; right to property; right to privacy and so on.

How much of this democracy has helped all Indians to have better incomes, a better life and better equality? Can it give these to all, even over a period?

Let us examine another softer issue: COMPETITION is considered to be good in politics (through elections and other means of propaganda, protests, public relations etc.) to ensure that people get the "best" leaders. Similarly competition between businesses is considered to be good to ensure economic efficiency and to offer better quality and price to consumers. India has seen many practical examples since 1991, when the economy was liberalized. One simple example: before 1991, telephones were the MONOPOLY of the Government of India. Only 50 lakhs telephone connections were there over a period of 44 years after independence. The elected leaders assumed that telephone is a luxury and not needed for ordinary people. Post-1991 thanks to the competition and private sector entry, now there are 96 crores telephone connections; they are increasing. Same can be told of availability of milk and many other commodities.

So competition is good; are there other softer issues related to it, which can affect society? So are many other words: EQUITY, INEQUALITY etc. each pointing out a certain softer issue. Also the word JUSTICE is used. After a major crime is reported, even before the alleged crime is investigated, media and vocal activists demand JUSTICE. Often times by this word "justice", the shrill voices desire IMMEDIATE PUNISHMENT of the person(s) who has (ve) allegedly committed the crime.

"Instant justice"appears to be the desire of persons who are affected. There are processes of law; they are perceived to be slow and are also slow in practice. So we need to look at the softer issues of JUDICIAL system.

Similarly the softer issues of what is meant by the QUALITY of education. Are issues of quality, ranking etc., easily measurable items with clear and quantifiable answers?

There are much softer issues of perception about various entities and processes of modern life. It is easy to say that all traders and businessmen are bad, greedy and are there to only fleece. "Middle men" are perceived to be criminals. Similarly there are many who perceive the statements of oversight agencies (we have addressed about these counter bureaucracies in one of the chapter) like CAG parliamentary committees or even court verdicts or newspaper reports as **ultimate "truths".** How much scope for errors exists in many of these processes (even without any intent of corruption)?

Lastly, is it possible to assume totally neutral media; or oversight agencies or even those who take executive action? Is it possible to "design a system where everything is decided by objective data, analysed by computers."Many persons tend to quote one study or a report of a committee and would like to base all decisions on that. Or try to find ultimate solutions in a magic institution like Lokpal?

We don't claim to have answers for all of these. Nor have we covered all items which can be called softer issues in the above illustrative narrative. Still we believe that all Indians should be aware of such softer issues, which become crucial for the modern world, as it is very many times more complex than it was for the lives of people even a century ago. There is a lot of information of all forms including disinformation misinformation and outright lies, (as explained by Murray Gell Mann; we have quoted him often) floating in the cyber world, which surface out through print media, audio visual media and also through the "minds of people" who are influenced by them.

For a better life, for a more peaceful and secure life, we all need to understand the softer issues, so that we can form our opinions in a more balanced way. That is crucial for the modern life; that is crucial for being in harmony with Nature, as it is understood during the 21st century.

Hence this chapter. We address some of softer issues to illustrate as to how we can approach each item when we are confronted with them directly or indirectly.

Democratic Systems

In the 21st century context, it will only be relevant to consider only the multi-party **electoral system** as DEMOCRACY. Even with this narrow definition, we find that there are several distortions; in many countries including in India, a single party dominates for a long time, a few decades or more; some of it naturally and some of it by what is called "rigging" as a process of COMPETITION by any means. The latter was prevalent in many parts of India, but mostly now eliminated due to electronic voting system and strong actions by the Central Election Commission since 1990's.

The universal adult suffrage in India having a huge illiterate population, has spread the base of democratic process. The identity politics based on caste, religion, language and even the charisma or nostalgia of old feudal and royal families have played a role in winning seats. Many caste groups who had been historically way down in social and economic hierarchy could come up with assembly/parliament seats. Then their "voices" are heard. Measured in such terms the Indian Democracy has done well. However starting with the Central Government in which a single party has ruled for many decades, a strong tendency towards dynastic rule has set in. The political party can be "captured" by a few individuals and their dynasties can be the perpetual rulers as the political parties have nearly monopolistic absolute authority to nominate candidates. The so called "inner party democracy" is more in terms of talking "nice things" and **there is no constitutional provision to enforce elections within political parties at periodic intervals; and to limit the terms of the office bearers.** Some parties have such rules. Most of them do not have or have it only on paper. Such an absolute opacity and immunity provided to the basic sources of or roots of electoral democracy have led to **several negative features in the functioning of the democratic politics** which **controls the governments** (in the State, centre and even in local bodies) and thus has become the very nature of GOVERNANCE in India.

First and foremost is the formation of dynasties in almost all parties, including those who came up with the slogans of inclusive politics. Along with **top layers of dynasties,** there are several **sub-layers of dynasties. Those Indians who were born in 21st century, that is, those who are now under 15 years of age have practically no chance of holding good position in political power hierarchy.** Such a possibility for all Indians is crucial for the **true spirit of democracy.** The current generation of Indian political leaders at the top who have no dynastic origins, were all born about 60 years ago!

But even in such parties, the Gen X is knocking at the doors and would like to use the "privileged access" of their parents or uncles or aunts or parents-in-law, to enter into high levels of the political ladder when they are young. There are many examples of such Gen X political leaders in India. **What happens to rest of the Indian boys and girls who do not have such an access? This is an important SOFTER ISSUE about the functioning of political parties in India, which are becoming family properties.**

Closely coupled with this, is the question of **funding of political parties.**

Even when there is a single "dictatorial" party, it needs large number of party members (workers); they need to be taken care of in terms of their personal lives, welfare of their families, etc. They would like to be better than the "average" people ruled by the party. That means the functioning of the parties requires huge sums of money, with of course the top and middle level leaders taking away bulk of such money for their own power and welfare.

In a multiparty system things are even tougher. **There is competition:** good one in terms of policies; but as it happens in India even now there are several bad parts such as "bribing the electorate", creating illegal hurdles for the opponents, which border on crimes (This is also described as "money and muscle power" by politicians themselves to attack the party which has won more seats).

In addition the "first pasts-the-pole" type of voting system for selection of the winner creates many anomalies; in multi-cornered contests a person **who polls just 20% of the votes polled or even less would be declared as a victor.** This system, has its merits especially that of simplicity, but also adds lots of uncertainties to the members of the political parties who aspire to win and to hold Government positions. Naturally it also adds to the anxieties of political leaders because there is always a possibility of losing their majority in seats and thus their opportunity to rule.

All these uncertainties lead to accumulating more resources to survive through "rainy days" (that is when not in power of Government) and keep fighting to win next time. **Thus fund raising has become the important game for the political parties; it is also seen as a mechanism for survival and perpetuation** (a Darwinian process). This is also another important reason why dynasties form within parties and hold on to it with death-like grip. Because the process of raising funds for parties is one of darkest, opaque and well-guarded secrets by all parties. What is declared for public consumption, is only a window dressing. Even parties with small number of seats raise such funds by use of various forms of blackmails: Extortions and "deals", as bigger parties do with greater ease. The darkness of the process allows for the individual actors who raise funds and spend money, to divert some to their own kitties! The higher up one is, the greater is such a chance. Hence there is the tendency to perpetuate their own leadership; as long as the party workers are kept happy through substantive crumbs available to them, they will not rise against the leadership. Power-struggles are left to the top echelons of the party. How such processes had built up from the time of Indian independence, is beautifully brought out in his fiction-autobiography "The Insider" by P.V. Narasimha Rao. It is interesting to read about the slogans of "land reforms" which are used for political gains but without actual implementation in order not to disturb the rich land lords. The current political activities around land acquisition are to be seen in the back drop early history depicted by P.V. Narasimha Rao.

The fund raising and expenses for the political parties for survival and victory, is the main source of corruption and black money in India. This is an open secret but talked about in public discourses and writings only sparingly. **But the fact remains that these processes within political parties without mandatory and well monitored democratic elections within parties etc., are important SOFTER ISSUES;** these have severely distorted the functioning of Indian governance at all levels from centre to the state to local bodies.

The resultant or corollary malignant softer issues associated with such a "democracy" are:

- Ubiquitous black money and corruption distorting all processes of economic reforms, economic growth and preventing the reaching of benefits to bulky base of India. It also leads to "crony capitalism"and "crony welfarism" (which is diverting of tax money for distribution to people with an aim of victory in elections, though such actions are very damaging to Indian economy and people)

- Parties are so self-centred that they look for every opportunity to disrupt the elected government from functioning. As the TRAI Chairman said in a recent statement about net-neutrality, **we notice daily a number of shrill voices, ably aided** by a number of media channels, to distort the process of democratic dialogue and rational governance. **Is it not the intention of the founders of the Constitution makers to use the legislative wing of the Republic comprising elected representatives of people as a medium of discourse of issues related to various aspects and arrive at a good compromise of interests between various groups?** But what is done is very different now. Block even good proposals. Replace dialogue and discussions with protests inside and outside parliament; disruption of civil life by various calls for bandhs (the State Governments being silent spectators) or use of dharna; fast-unto-death protests.

The electronic media has a number of opportunities to increase their viewer rating through shrill shouts in the panels (no discussions! !).

The real function of the fourth estate (Press) being a forum of good public discussion of various issues (pros and cons) to educate the elected representatives and party leaders at various levels, has now been reduced to the pressures to be alive in the market economy (emphasis is therefore on advertisement of various sorts to tendentious reporting to "paid news").

The emerging fifth estate, that is, the social media in India is still dominated by shrill middle class fads and fictions though there are a number of good features. These shrill voices again spill over to electronic media and print.

How much of these processes are "manipulated" by big business and foreign interests is a serious SOFTER ISSUE. The points are occasionally raised by political parties to "score a mark" against rivals. But it appears that the political parties are closely linked with these big businesses and foreign interests as they are easy source of funds.

Thus the Indian democracy has many "soft bellies". The reason we are raising these SOFTER issues and **pointing out the areas of soft bellies,** is because its functioning has now reached a state where **governance of India is becoming very difficult, if not impossible. Indian economic growth and therefore possibility of increasing incomes of all Indians are at stake.** Under such populist pulls, Indian agriculture may be affected very seriously without any possibility of the needed investments, technology, supply chain linkages and human skills. All those marginal and small farmers and the landless rural people, will be forced to leave villages as rural primary sectors do not provide the much needed opportunities for earning a decent income. Traditional artisanal activities cannot provide good incomes without the technology and design inputs to suit the modern world; such traditional artisans will sink into further abysmal levels of subsistence. The "free food", "free medicines", etc., promised through economically and financially unsustainable Acts of Parliament or State Legislatures will only be a mockery of the process of Law. How can government bear the costs of such operations without additional tax, resources of tax? Whom will they tax? An increased tax on businesses will drive them away from India to other destinations like China, Vietnam, Sri Lanka,

Malaysia, Thailand etc. The middle class (even the fragile upper middle class) cannot take the burden anymore. They will only do more protests, often with single point agenda: Sack scam officers, enact a bill, bring justice, have a CBI enquiry to expose the scam amongst a huge hue and cry. Those who have to do serious governance will be igniting one fire or another; often cynically hoping that some bigger fire will make "people's" mind shift to another fire and so on.

Serious actions in the short – and medium – term are forgotten. Occasional palliatives are a world level Indian Victory in Sports or an Indian NRI being recognized in USA; IMF, World Bank or some study reports that project India being great by 2030 or 2050.

Copying the Developed West

There is a group of elites who are hell bent on bringing Western standards to all of India, conveniently forgetting the fact that those countries evolved over centuries; also those countries have their normal distribution on many items be it for "quality" in schools, colleges. Not all are Harvard, Yale, Stanford or MIT.

For a country like India with Rich and Super Rich and Upper Middle Class persons being only about 300 million, with a Bulky Base of 700 million persons surviving at low subsistence levels with every next day being an uncertainty and about 200 million middle and lower class Indians just gotten out of their subsistence levels (due to some higher education or city life etc.), it is impossible to impose Western country standards for education, or health care or pollution levels or various forms of human rights or welfare guarantees by law etc.

For example, the higher education system is being skewed very heavily to copy the Western standards of citations or ranking. This creates a serious problem of reaching reasonable education to all Indians especially the necessary economic and social skills required for the Bulky Base and Middle and Lower Middle Indians, to get high value incomes.

Even during the independence movement, there were some persons who considered themselves modern and progressive, took every opportunity to deride things of India; they would like to have the British/European viewpoints to prevail over things of Indian origin. They would like to negate good things of Indian past through their (only type of) "scientific history". Only after Western scholars bring out facts on India, then they reluctantly agree. This tendency continued. With the economic opportunities provided after Independence such scholars "captured" the Indian elite transactions – universities, economic institutions, social institutions law making etc. With increased globalization, such elite groups increased in number. Now they would like to enter into making standards. Let us quote some portions from the recent 24 March 2015 issue of The Economist. "Top of the Class: Rankings". We are specially quoting this because Indian educational administrators are now obsessed with citation index and rankings. And vested interests in foreign countries and India are making a frenzy of it. Even courts are swayed by these writings, perhaps enamored by the visions of "equality" in constitution. Very rarely contrarian view in the Indian context (even mildly) is allowed to appear in press or official papers. Even such contrarian views appearing in serious Western media like The Economist are sidelined. Let us see what is said in the article:

"There are plenty of worries about the effect of rankings. Baharam Bekhradnia, President of Britain's Higher Education Policy Institute, reckons that "they are worse than useless. They are positively dangerous……."

"One concern is that these metrics measure inputs rather than outputs. "The indicators are resource intensive. They are about wealth", say Professor Hazalkorn. But the main objection is that most of the metrics, directly or indirectly, concern research. There are no good internationally comparable measures of teaching quality. So one of Mr. Salmi's favorite universities, the Franklin W. Olin College of Engineering in Massachusetts, which he says "provides a superb learning experience to its students" does not feature in international rankings because it does no research".

In USA which allows diversity to flourish such institutions can flourish. But in India obsessed with unitary control of education etc., from Delhi (through UGC, AICTE, NAAC, etc.) a few of such institutions which exist in India will be choked of project funding from Government sources. The academics and science bosses (who are again a coterie with a narrow base controlling many of the policies and funding), would crush these institutions still existing in India in several ways; the courts and media may amply help them!

Let us end with quote from The Economist article:

"Justin Lin, a former Chief Economist at the World Bank and currently Director of the China Centre for Economic Research at Peking University, has a habit of swimming against the tide. In 1979, he defected from the Taiwanese army to China, swimming across the narrow strait from Taiwanese-administered Kinmen to the main land. These days his contrarian nature has tamer outlets: he doubts that China should be in the race to create world class universities if the concept is defined by the number of its faculty's publications in journals dominated by the West's research agenda. "Who cares about world class research if it doesn't apply to the conditions that you are in", he asks.

How many, in India, are asking about relevance (in the Indian context) of many issues they are propagating for India?

We will continue to elaborate on some more of the **softer issues** in the following part of this chapter, as these are very serious and crucial for several millions of Indians who may not be aware of them and may be swayed by populist slogans, which are raised by self-centered narrow power-bases in India.

Indian democracy still provides a forum, despite several dangers to it as pointed out to raise such issues and find India relevant solutions.

We have addressed a few softer issues in some detail, such as the elements concerning democracy and its effect on governance. We had also mentioned several words which are commonly used like JUSTICE, EQUALITY etc. which have many imbedded softer issues; but we had not yet dealt with them in some detail. One item we had ended the last article was about Indians copying the systems from some elements of the western developed countries. We had dealt in detail the obsession with citation, ranking etc., especially in higher education. We need to address in some detail, how such a copying of the West is impacting a few other vital sectors.

In School Education, there are many such examples. One glaring one is the Right to Education (RTE) Act; the political system and a number of vocal activists had **identified the passing of the Act as almost the end point of their transferring the constitutional rights to equality in education to all Indian children.** Several features in it such as passing all students up to the eighth standard without holding them in the same class for poor performance etc., have further diluted the effectiveness of delivering good education to the children born in the Bulky Base (BB).

The focus of giving special attention to the learning by first generation learners, and several million children who have various family and other socio-economic, constraints in learning, has been forgotten (such a focus was poor even before the RTE Act). Instead, after the Act, the overarching activities of the State governments have been to identify the mechanical details of the class room size, quality of building, teacher-student ratios (irrespective whether they teach or not!), and identification of 25% poor students to get admitted etc. The only indicator of gross enrolment ratio (GER) overlooks the requirement of effective delivery of teaching; the drop outs, etc. In between lots of funds are spent on IT based education which are major issues for the developed world which have uninterrupted electric power for almost all their citizens; very high internet penetration etc. So, much of the funds are diverted for these buzz words:

- Universal access of education as a RIGHT
- Enabling equal access of quality education to all Indian children.
- Inclusive education.
- Bridging Digital Divide etc.

But the delivery of good affordable education to the children in the Bulky Base is submerged under these famous words. We don't know when the National Knowledge Network (NKN) is going to benefit youth and others by giving the Rust collar/Blue collar/Grey collar/While collar skills.

The elite and media are happy in reporting that only 15 to 20% Indian students with graduate degrees are employable. They use the Western type survey, mostly limited the questionnaires to the elite employers of MNC's and other big companies.

We should all learn to penetrate these buzz words/slogans and see the softer issues underlying the affordable relevant and good education and skills. Words and Acts do not provide them.

The softer issues and realities behind

- Equality in education
- Inclusive education
- World class quality education etc.,

Require a deeper examination.

Language of instruction and/or inclusion of local languages as a compulsory subject, are other softer issues, mixed with high emotions.

One softer issue is that what is the status of Indian languages to deliver modern knowledge and skills to

our children? Do we have enough good quality text books and their continual updates?

The issue of languages is raised politically as perhaps a reaction to the copying of West, which is done by the elite and political classes themselves. "Run with the hares and run with hounds" appears to the approach of political parties. Let us look at the result of such a copying of the West on the Medical access for our citizens.

Health Care System, softer issues of copying the West – If one studies the Government policies and actions as well as the media build up, over the past two decades since liberalization, the term "quality medical care" has been made synonymous with the Corporate run hospitals. Government run systems are systematically shown in very poor light to be much worse than they actually are. Medical insurance companies are focusing on the treatment by these higher and private (giant) hospitals. Even many government medical schemes include such private hospitals in their approved list, because they would like to ensure that the upper echelons of Government servants, politicians (who are MP's MLA's, Ministers etc.) can avail of such "world class quality medical care". Even during their jail period VIP's are specially treated in such private hospitals. **The Government schemes also allow higher levels-of-the powers-that-be to go abroad for treatment, paid by Government with tax payers' money.**

As a sop to Indian Government sector hospitals, promises are made to build high quality AIIMS like hospitals in many States. The State governments compete "madly" to have one AIIMS located in their States.

While there are many positive features in the medical care systems of the developed world, it is also true that they have become very costly. With increasing "protocolization" of medical practice (that is, not allowing the clinical physician to make intuitive decision; the prescribed manuals are almost like industrial steps – with many medical tests, often repeated). There are also now attempts to develop algorithms to analyse such massive data to prescribe course of clinical action.

In addition, medical service is being treated like any other consumer service and the doctors/hospitals are subject to litigations. In USA, the insurance against such litigations as well as costly tests (again to protect the profession by citing "objective" data and due diligence in courts) have risen the costs of health care to levels so prohibitive for about 20 to 30% Americans, in a country which has a GDP per capita about 15 times more than global average GDP per capita.

Indians, the bulk of whom form a huge part of the global poor, have very poor access even to an ordinary doctor. We have written about such access through Mobile Hospital in chapter 21of the earlier book. But such items do not catch the attention of the powers- that- be, as there is no big money in those schemes. They cannot be called "world class" though they will serve Indians very well as has been proven in Uttarakhand for more than a decade.

The real softer issue in medicare/health care policies/ actions is that bulk of the investments in India in this sector are towards the upper end Western style private hospitals. Incidentally the costly biomedical equipment are about 100% imported. No Make in India there! The access costs for these hospitals are prohibitively high for most Indians. Those who can afford are Rich/Super Rich and some Upper Middle Class (UMC) Indians whose medical care costs are borne by their companies. Again, policy attention is given to reduce taxes on insurance cover for treatment in such institutions. All these are orchestrated so well, as if all the "world class quality" health care is given to all Indians; in addition there are considerable media hypes that these hospitals will create health tourism industry and lots of jobs for India! As a small token, a few new AIIMS's which are very capital intensive are also budgeted. While more government hospitals are welcome, the way they are run, nobody bothers; good doctors do not join. These hospitals languish with overly excessive crowds. **Because over a period, this soft push of copying the West (albeit for use by a small minority of Indians but presented with a public relations exercise as if it covers all Indians), has eliminated many simple General physicians and small affordable nursing homes out of existence.**

The only thing, the Indian health care system has done well is in vaccination/immunization programmes. But in addition, people require continual reference to doctors, when they feel ill.

We won't be surprised if more and more laws will be passed to weed out "poor quality" health care; "the quacks"all in the "interest of" and "in the name of" protecting the health of poor Indians. Social media may be used to make a few anecdotal examples to pressurise governments to enforce such laws; to force medical stores not to give medicines without a qualified doctor's prescription etc.

In between, some "stringent" actions will be announced against the private sector hospitals to treat certain percentage of poor Indians (say 25% as in RTE Act) in their hospitals. Note that **it is not 25% of all poor Indians who desire access to hospitals, but 25% of the capacity of the private hospitals.** There will be legal bureaucratic wrangles as to how much the private hospital can or will do: "we will give totally free consultancy of the top doctor; our equipment time will be free; but consumables the patients have to pay,here we are ready to bear 50% of the cost etc.". Even a fraction of these charges will be well above what a poor Indian can pay. Add on to it the travel costs of the patients and relatives to such places, cost of accommodation among all the other expenses incurred.

We have dealt with Education and Health care in some detail, **in terms of softer issues of copying the West** is because of the grave danger staring at us: Most Indians cannot have affordable access to health care in the near future. Even many middle class Indians (let alone those in the lower middle or Bulky Base) cannot go to such costly hospitals. Functioning Government hospitals, (too few in number for the huge number of population) will collapse in operations due to the rush beyond what they can handle; they may face public anger, protest etc.

Therefore let us wake up and look at the softer issues behind the slogans of "World class quality health care" to all Indians! Let us give them at least some minimum affordable access **now and in the near future!**

Other Examples

There are plenty of other examples of copying the West. They are pushed into the minds of Indians (even poor ones) through high pressure advertisements in all languages: from facial creams to packaged foods to many other consumer products. Well we can leave them alone now as market forces will take care. A young poor woman who cannot afford a costly facial cream would not buy it or find some alternates to satisfy herself. So let us not open many fronts. Also these products provide employment for Indians (from manufacturer to retail).

But affordable health care access and education for all are crucial, as they affect all Indians in a critical way, also jeopardizing their future existence. We request the readers to address these two sectors (Health care and education) very carefully for softer issues while copying the western models.

Justice System

Most Indians look for the "final" JUSTICE in the Apex Courts. Many believe that apex court judges are the real saviors from the corrupt politicians, bureaucrats and the business persons. The way media reports are done, that this or that court "turned" down government proposal or this or that Act, as if the courts (especially the apex ones) are the tables in which the "buck stops" as if they are the ultimate bosses. The Indian Constitution is much more nuanced, following the British traditions. There is no superior body between the three wings of the State: the Legislative, the Executive and the Judiciary. Whether a legal provision enacted by the Legislative or implemented by the Executive is in accordance with the provisions of the constitution, is what the Apex Court decides. In the British tradition, the legislative or executive may absorb the observations and ruling by the Judiciary, and modify their Acts or implementation accordingly. They have their right not to accept as well.

It is necessary for all the three wings, the political parties, the bureaucracy (the administration) and the media including elite public opinion makers not to disturb the delicate balance; not to encourage any of them to cross the Lakshman Rekhas, which are very delicately marked by our wise founding fathers of the Constitution. President of the Indian republic is not like the President of USA. He (she) is not a vetoing authority. His (her) inputs are much more subtle to all the three Wings. So is the role of the Prime Minister of the Indian Republic, who has much more powers vested in him (or her). But he (she) does it through the Cabinet, in which he (she) is the first among the equals. These subtleties are presently brushed away. Especially since the turn of the century when telecom revolution was picking up and during the late 1980's and 1990's when the TV Channel revolution (through Cable TV's) was picking up, the media (electronic and print) have picked up the methods of populist (spicy) American media. They are in search of instant JUSTICE and a powerful MACHO LEADER. Icons are created, many persons "fall in the trap", only to fall out of favour a few months or years later.

Unfortunately, one is pained to see that some members of the JUDICIARY also fall prey to such populism: "Green Judge" "Human Rights Judge" etc. Despite this, in the overall the Indian JUDICIARY has kept its balance. But whether they are overreaching is a softer issue, which the JUDICIARY and the legal community has to ponder, especially within the framework of the Indian Constitution.

Though the Indian Constitution specifically provides for consideration by the Indian Legislative of the Acts/Legal Positions taken internationally (in the democratic nations), there is a growing tendency in the Indian Judiciary to accept the "international" provisions without due deliberations by the Indian Legislative. It is the duty of the Judiciary to raise these points to the Indian Legislative and Executive to take positions, before they adopt them.

Another totally different but important softer issue that we all need to be aware of is whether there is an "ultimate single truth" which the JUDICIARY can uncover.

If one looks at the judgments in the lower courts, special courts being turned over completely by the High Courts and later in the Apex courts, very differently, any rational persons has to come to a conclusion that "truth" uncovered by the courts is the result of many complex processes of law, very narrowly defined. The famous dialogue of Annadurai (beloved Anna for the Tamils), in one of his script for a Tamil movie

"Chattamenbathoriruttarai

Athilvakkeelinvaadamoruvi Lakku"

"Law (Justice) is a dark room. In it, a lawyer's arguments is a lamp".

Another (generally positive) fact is that the Judiciary is affected by the socio-economic millieu and the developments in the private and public lives of people.

If one studies the rulings of the courts on labour relations, industrial disputes, or private lives of people like marriage or many such items, one could see almost revolutionary changes in the outlook of JUDICIARY over the past over six decades in India. In the 1950's or 1960's or even later, the court judgments will be in favour of the labour unions (what is now considered as the labour aristocracy) or actions by public sector units, keeping up with "socialist" policy. Similarly in the private lives of people, monogamy and strict laws on marriage were the preferred as the Laws enacted by the British, which are a part of "constitutional provisions". Divorce was made very difficult. Live-in relations would not be accepted at all.

By the turn of the century all these are different. Labour is more a contract, not a lifelong right. So are marriages and even live-in relations. In this transformation of the judiciary, copying of the current Western traditions (American and British) has played a great role.

With environmental activism and human right activism in the west, what was considered as "public good" has changed in the "eyes" of Indian JUDICIARY? During the 1950's, 1960's, 1970's and even during 1980's, Government could acquire any amount of land for establishing military (defence) establishments, government industries/institutions, or dams. Many persons who had been displaced would be "given" a compensation" as decided by Government. (Executive). It was difficult to win a case against Government action.

Now well into the 21st century, it is almost impossible. The JUDICIARY's balancing of the "public good" versus "private rights" or "human rights" or "environmental good", is very different. **These perceptions are not often written into the books on law or in Indian constitutions.** These are the "judgments" of what is closer to "truths" amidst conflicting requirements of the modern Indian society.

These are also strongly influenced by what is happening in other "democratic" societies of Europe, US, Australia and others, mostly English speaking as our JUDICIARY, elite, media persons, etc. **strongly anglophile.**

We have elaborated on a few of these points, mainly to high light that there is **no ultimate JUDICIAL truth.**

I have explored some of the facets what TRUTH is in several spheres of human activity: LAW, JUSTICE, SCIENCE, ARTS in my book "In Pursuit of Happiness" published by Rajpal and Sons, New Delhi, (2006). It was result of my earlier intense study about Human Knowledge (as it is currently understood in the modern world) and Happiness for delivering R.B. Datar Memorial Lecture 2002 organised by the Supreme Court Advocates – on – Record Association (29th November 2002) at Speakers Hall, Constitution Club, Rafi Marg, New Delhi. I had identified different knowledge bases and their relation to the Maslow needs:"Search for truth in arts, science, religion, law, economics etc., is a creative process with special features of self-actualization……"

I have described them in some detail for each area of knowledge. For the legal systems: "It is found that Legal systems definitely can play a role in the safety part of the Maslow needs. Can they play a role in controlling evil elements of those who disturb the esteem? In the civilized world one cannot resort to Dharmayuddha with swords. Its like a metaphor that legal systems are the swords of the modern societies-they can be used to sublimate physical aggression into apparently non-violent forms………… can the legal and judicial systems help regulate the needs of esteem into socially useful channels?".

My study showed (shows) that bulk of the humanity is happy with meeting the lower levels of Maslow needs. Their needs for self-esteem are limited to being respected and loved by the near and dear ones around them. **But there are a**

small percentage of persons who would like their self-esteem extended far beyond: they are those who want power, fame, influence etc. But many of them may want to use their knowledge, access to knowledge etc., to control others. The more the people are under them, they feel better. In the current day parlance to have followers in Twitter in millions should satisfy some people. They even would like to compete in this as to who is top in the world.

But such a direction of self-esteem to control others can turn out to be dangerous to others in the society.

The basic principle of the modern judicial systems in democratic societies, is to ensure protection of ordinary persons from the "powerful" ones; be they politicians, rich persons, elites, etc. **Do Indian judicial systems fulfill that basic principle?**

It is a big question for which the answers will be mixed.

- Ask many thousands of people who are engaged with courts in some way or another: for a property dispute; or being arrested by police, presented to the court to seek relief; or fighting with the employer (e.g. Government) to get compensation etc. Their views on the performance of judiciary will be very dim.
- First and foremost, the courts are not easily accessible; one needs to engage a lawyer/advocate through whom one has to approach the Court; so many affidavits; signatures of Notary; etc. The strange situation is that even many well educated persons would not understand this language: legalese. Also in the interest of access to people, local languages are used; many outsiders have to believe what the lawyer says; for the illiterate/ semiliterate persons, the fact of its being written in their language gives little comfort (Also the final legal

provisions in case of interpretations of law "in case of conflict, the English version of the law will prevail").

Thus, a huge hiatus is built up between the courts and the people. Procedures are archaic: from getting of stamp paper to giving many such "submissions". Another big barrier is that the person who gives "Surety" has to be **a gazette officer or have property.** Most Indians are thus excluded.

It is therefore no wonder that many tribes had lost their rights to land since Independence through such legal processes. The compensation for displaced persons or accident victims etc., does not reach them for years. **It is not just due to the "bureaucracy" as is commonly made out. The legal processes involved in proving the identity of the person(s) affected, surety for such affidavits etc., are long drawn processes.** They are fertile lands for creating middle persons and touts. **All in the name of rule of law! This entire process of functioning of the input-output systems for the lower courts, intermediate courts and apex courts needs to be examined.** Keeping on generating more and more documents etc. is not application of mind or the impartial delivery of justice. This softer issue of input-output processes to and from courts is a serious one, causing inordinate delays leading to "justice denied".

- The courts appear to act in a way that time stands still, the moment the case is admitted. The affected persons can wait eternally. Getting orders after a few years or decades even to clear one's name of an alleged crime has terrible consequences on individuals, their families and even children (who lose their careers).
- Another tendency which is emerging more often now is to pronounce judgment on events which have occurred many years ago on points of technicalities of laws: the order on coal mine allocation is an example, whatever

may be the merit in law. We are seeing a number of cases in Bangalore regarding illegality of housings, or commercial complexes etc. based on activist applications, which take several years to be heard etc. Lakes or forests are encroached or wrongly allotted by Government decades ago; these properties exist on them; owners have "valid" government approvals (for decades), receipts of regular payment of various taxes, electricity bills etc. ; still these are ordered to be recovered. In such cases demolitions are done, even though after demotion lakes are unlikely to rejuvenate! Are they going to recreate the lakes?

- Big question is who is the real culprit? Who in the government gave the approvals? How will a citizen know that a particular designated Government Office is wrongly allotting a land? When the procedure of getting a land totally "legally" and getting various certificates is a Himalayan effort, would a same person ask each Government official who signs/stamps as to whether the government approval is legal?
- On the other hand those who are rich and powerful can engage a fleet of highly paid lawyers/advocates to argue the cases from one court to another. They can safely postpone most cases (which are even blatantly illegal) to years till they die after enjoying all the benefits of money and power.

A few cases of arrests (very small in number) of high visibility cases (from Delhi) or of a few politicians under media activism and court directions, do not disprove the fact most of the cases of the powerful and rich drag on through technicalities and often let free (often through fading away from public memory).

- In a recent piece of "A thought for today" (1 May 2015) Times of India, it was given "The power of the lawyer is in the uncertainty of law".

Another serious issue which is overlooked is the tendency of some courts to give directions on matters which are purely technical. Issues which involve complex interplay of technology, engineering, economics, business considerations etc., are handled with some single point agenda. One example is how Delhi was rushed into use of CNG in Delhi's auto rickshaws, taxis etc. Did it solve Delhi's problems? Now when there are shrill voices about "the dirtiest city of India", Delhi, do people ask the question whatever happened to CNG? Are there any considered study that "guarantees" that banning of commercial taxis older than 10 years will solve Delhi's problems? What about trucks, buses and diesel generators of the rich and powerful? Also atmosphere is not static. **What are all the sources of pollution?** In between there will be green activists wanting to enforce Euro 4 standards for all Indian automobiles. As we have pointed in a chapter on **Standards,** there are many geo-commercial forces at work. Euro 4 standard will make most of the current automobile manufacturers "obsolete". They have bought technology paying heavy price and have mastered some elements of it, to manufacture in India and even export some. The sure way to "kill them" is to ensure standards which will be costlier for them. Also many middle class Indians have to for ego the desire to have a car at a young age to go to work.

There are now slow but steady campaigns to introduce standards in Indian food and food supplement Industry to have European/USA standards. If enacted and enforced many small Industries will collapse. We pointed out this matter because some courts seem to be attracted to western standards and would like to give a ruling and direction to have them in India through specific rulings on products: Cars more than 10 years is in such a category. Does any of them bother to see how most citizens of Delhi live? In slums, low cost housing colonies, and even in middle class houses, barring those areas where the rich and the powerful live. Their houses have much more pollution.

To invoke a slogan of "Right to fresh air" is a mockery about the Bulk of India which struggles to live. Most rural households use biomass burning for cooking; even in towns and cities.

It will be in the interest of the people of India, if a balanced approach is taken. Let us look at the evolutionary processes of economic growth. India cannot aspire to be like Sweden or Switzerland or Germany of today without understanding the processes they had gone through. Let such issues be approached with caution. Let our judicial systems accelerate their work on other areas of pressing need for bulk of Indians by dispensing speedier justice and let them have their dues paid without unnecessarily complex systems of judicial processes as pointed out about earlier.

It is essential to discuss the softer issues regarding the JUDICIARY system (not so much as the appointment of judges – which also has many soft and subtle issues!) to find simplified solutions to make millions of poor, lower middle class, even middle and upper middle class Indians to feel safe about our JUDICIAL system.

Presently most of them have given up on that though they are struggling through various litigations & courts, as they have no other option.

It is necessary for the judges, the lawyers, advocates, administrators and other influential persons to come down to the realities of ordinary Indians instead of trying to compete with the west by enacting the most "progressive laws" which may have meaning for them in their social-economic-evolutionary phase.

Otherwise we will find road rages, instant justices delivered by mobs of persons, khap panchayats of various forms, etc. Worst still will be a growing tendency of people looking for **safety to form their own small** communal/group networks (based on caste, religion, language or geography, etc.) and enforce "law" as they deem fit to protect their groups from "injustices" done by others.

Yes, people will take law into their hands, if the LEGAL/ JUDICIARY system does not lend them protecting hands **in time.** In a democratic society and with increasing speed of communication systems, the judiciary/legal systems cannot claim to be impartial only through blind folding to ground realities of ordinary people.

What Nature can offer in the modern world, can be enjoyed by all, only when the survival instincts of people are risen well above their primordial state, to a highly civilised state through responsive legal, administrative judiciary and law-enforcing systems.

Unlike the earlier species which occupied the Earth and also unlike the early human beings, the human beings in the 21st century possess extremely high destructive capabilities, as we witness around the world, in developed countries and developing countries including India.

Hence we need to address these issues early and speedily. The desire for safety, security and justice are one of inherent Maslow need for all human beings.

❐

20

Competition and Interdependence: Capital, Technology and Society

Indian media through its penchant for hyping many local issues beyond proportion might have created a clumsy cocoon around most middle class Indians, insulating them from many serious happenings around the world. TV Media, which ought to have been much more capable to cover the world news, is still worse.

What is happening in the Middle East countries is little known to average Indian watcher of TV or newspaper reader (in English or in any other Indian language), though they are likely to have far reaching impact on India. Indians are given only news about abduction of some Indians there; a week passes in sensationalising the issue. There is practically no coverage of Africa, Latin America and even Europe, Russia, China, Japan and South East Asian countries. Australia comes into horizon for reports of some Indian students in trouble. Prime Minister's visits to some of these countries, the latest being to Iran (2016), do not receive special coverage.

Even much of the economic and strategic gains from the foreign visits of Prime Minister, are trivialised in reporting them as a part of divisive local domestic politics ranging from the dress he wore or such trivia.

Even some laudable sport successes like that of Sania Mirza winning in Wimbeldon are missed because of obsession with another salacious murder case; as also the successful entry of Indian women hockey team into Olympics after 20 years. Many rapid changes in Chinese economy ranging from successive yuan devaluations to stock market crashes are not reported well, most often ignored except for reporting Indian SENSEX reactions. Print media is better in that at least there is some coverage; one or two articles come with some analysis.

We may add many more illustrations. In the overall, most Indians (professors from elite educational and scientific and technological (S&T) institutions, many hi-tech IT/ICT professionals, journalists, artists, writers, "intellectuals" of various sorts, bankers, politicians, activists, NGOs etc.) are only obsessed with the very narrow issues relating to their surroundings or pet subjects or fads. Even those interests hop from one to another: anti-corruption crusade; onion prices; some violent incidents like murder, rape, some sport etc. Even those who invest in stock markets have poor research back up; that is one of the reasons why SENSEX tends to be volatile. May be some "invisible hands" are taking advantage of the general ignorance of most of the retail investors in stocks!

Most Indians perhaps think that is enough. There are governments (Central & States) to do things and often to be used as "whipping boys/girls" for media, protestors, "opposition", ban-seekers, etc. They perhaps assume that such "vibrant democracy" will take care of things. India is any way destined to become a "super power" because of the youthful human resources. Some may think that our culture and civilisational heritage will take us through any challenges. Or India's "soft power" will bring guaranteed success.

The reality is that the current day world is globally connected and COMPETITION and INTERDEPENDENCE (the two opposite things) are the driving forces of most of the economic, social and even political aspects of human beings of the world (which ever country they live in) has not dawned on most of Indians. Understanding such a reality is crucial for the present and the future of all Indians; if understood well it would change the attitude towards economic, social and political issues. If more and more Indians know about such realities and desire to adapt to meet such realities, then the politicians will also follow quickly. They will understand that the older forms of divisive policies for intermediate or short term gains will not be accepted by people of India. Media might change too!

Let us understand the role of competition and interdependence and their impact on CAPITAL, TECHNOLOGY and SOCIETY.

Let us first place the position of capital, labour and society in the context of India and the world @2016.

Capital

The early origins of capital came about when kings and emperors minted coins. These coins became a convenient tool for paying for goods (which were bartered before) and for the labour. Kings rewarded scholars and poets with gold coins. They could use it to buy land or house or other goods. Over a period, traders and a few rich nobles (who were in control of large tracts of land and labour) started accumulating money (coins) far beyond their daily needs. They used it for lending to others who were short of money and the lender did not give this free; there was a payment of interest. This is where money started multiplying more without doing an agricultural production or an artisanal product or rendering some service like ploughing the fields, harvesting or doing tasks of barber for personal hygiene or paying a local *vaidya* who gave medicines.

Similarly, the kings used the method of collection of taxes which were in terms of share in agricultural produce or in coins. Kings also accumulated wealth; but kings used them to build palaces, public places, to raise army, to do governance in addition of course their personal enjoyment. The money was more for getting products and services. It is the money lenders who gave money a special status of CAPITAL which can reproduce itself; it was not a win-win situation. The lender went on accumulating more and more money (he is to take care of it from defaulters and more importantly robbers). It is this accumulation of **capital** in itself created rich (and in contrast poor) people. Rich traders also extended their domain to other nearby countries (within what is the current day India as there were many kingdoms) and also outside to Asia, Arabia or China.

So capital accumulation led to growth of other economic activities and better lives as many others could consume better. In countries like the current UK, it also led to the Industrial Revolution. Persons who had surplus money beyond day-to-day consumption needs could build machines, ships etc. They could hire a large number of paid workers (LABOUR). Those with Capital used the LABOUR; produced more products than consumed in that local society or even kingdom. They created more surpluses of money (capital) through their local and international trade. Kings, of course, collected taxes and also provided security for those with Capital. For the neighbouring countries, kings used the show of force in order to gain access to trade and thus grow the kingdom's Capital.

But still the concept of CAPITAL was mostly tied to the asset base of the manufacturer (building, machines, raw material) and the labour force they employed. The technology was imbedded in the machines which used thermodynamic or electro dynamic power in addition to manual labour. The productivity of the

LABOUR (the human person) was increased manifold by these machines. Ships carried many goods which were several times more in weight and volume than it was possible with manual power based boats or animal based land transport.

In this process there was growth of capital, technology and labour (who had to be much more skilled to use the new technology-intensive-machines and tools). The societies had more wealth than they ever had before. There were also victors and vanquished. Many wars were a part of such growth. Technologies also gave birth to new machines of destruction; the knowledge of the many well trained scientists, engineers, and other professionals like geographers, historians gave new war strategies and tactics. There were organised military personnel (new forms of LABOUR) using technologies and strategies. CAPITAL owners invested on military-industry as well. The combination of CAPITAL-labour-technology helped the rulers to win wars; to defend and to threaten other countries so that their own country (society) can get a better deal.

Freedom for Capital from Assets

Still, even through all these developments of the past, upto about three fourths of the 20th century the capital could be identified with assets;such as the land, machines, labour force, raw materials and technology (intellectual property rights – IPR's).

With the rapid growth of computers, communications and information technology (ICT), and also the new modern globalization which started around 1980's giving dominance to market economics, world trade was transformed. Associated international regulatory regimes for protecting such a global trade came through World Trade Organisation (WTO); for the first time Technology also was brought under the WTO regime. Developed countries not only had control of lots of CAPITAL and also possessed IPR's for most of the technologies. They could thus dominate the global trade which was in theory "free" – free access, "market driven" etc. But those who had capital and technology won in competition.

The production chain was split into small elements of the value chains. The developing countries which had cheap labour, cheap land cost etc., provided the arbitrage for the CAPITAL investors by using their facilities to produce some out sourced goods. Thus the foreign investors preferred outsourcing to places like China, India, Taiwan, Malaysia, Thailand and many other developing countries selectively. The local CAPITALISTS would set up the necessary facilities and engage trained labour (and get them trained by the foreign technology giver); they supply to the specifications of the foreign principal business. This was so, for most manufacturing sectors. For the services sector, ICT was of great help to provide such outsourced services (with value addition) to foreign companies in USA, Europe etc.

CAPITAL could be transferred from one country to another; from one company to another at electronic speeds. This gave CAPITAL a special position which is not there for any other good. Global Capital flows grew in leaps and bounds with financial services emerging as a profitable business of its own (in principle it is the same old method of money lending and charging interests). It covered many other items like insurance (not just life but many of forms of risks) and derivatives (which are a form of risk coverage with complex mathematical models, which can sometimes lose touch with realities of life).

CAPITAL servicing related business and many other IT enabled services acquired a new position of market evaluation – market capitalization with very little connection with actual assets as was calculated earlier.

The capitalists who promote businesses increased the capital base of the companies they "own" (control), by distributing retail shares on a large scale: Initial Public Offer (IPO) and later sold through "stock exchanges". In the stock exchanges the trading of the shares are so unrealistically different from the asset base of the companies. Thus trading in stock exchanges has become an autonomous activity by itself. It is CAPITAL, CAPITAL-all over, with totally different effect on the economy. It has little relation to real GDP except that the wealth created by such trading and other financial services, is shown as economic growth. **That is the reason we notice jobless growth.** Playing around with capital has become a "wealth creating" economic activity for the capitalists, bankers, and the retail investors (millions of middle class persons). When markets "crash" the loss of capital which is reported in crores of rupees, is much more of the middle class investors. We have seen in India, now in China.

But this new form of CAPITAL is competitive (companies do many things to increase their market capitalisation i.e. sale prices of equities in stock exchanges) and also highly interdependent on world capital markets. When markets go down in some stock exchange, many others follow suit. If monetary policies are changed in capital – wise-powerful countries, stock exchanges have immediate reaction (at electronic speeds).

Technology

For global competition in real goods and services, availability of cheap labour and land, and lower taxes alone are not enough. Some elements financial engineering and marketing techniques are important to give a lower price and "feel good" factor to the consumer. Competitiveness is dependent on consumer satisfaction as well. Thus the capitalist, manufacturers or service providers and the consumers are interdependent.

Role of technology in enhancing competitiveness is to reduce labour employed (called as productivity increase) for a given output; reduction of raw material and other resources like water, electricity, energy(called efficiency). In terms of consumer satisfaction; increased ease of operation by the consumer (user friendliness); appeal to emotional aspects of the consumer (aesthetics; ergonomics). The latter two features come about usually through an element of engineering/ technological process called DESIGN. Design has emerged as a multi-billion dollar business, and a service industry of its own. DESIGNS form a part of IPR's.

Again all technological processes in the current day world including design are highly interdependent. One cannot be self-sufficient in all the technologies (and IPRs) and designs. Many companies from different countries will be involved. Sometimes the experts (high level labour) from different countries may be required to provide a competitive edge. For example, designs from Italy have an edge; many companies outsource to them. Engineering from Germany often in terms of finished machines (called embodied technologies) is used. For example, textile industries in India are strongly dependent on sophisticated machines from Germany and Korea for having their competitive edge in global markets. If one visits Tiruppur textile companies, one can see this reality.

Without heavy imports, the Indian economy is not able to do high tech (IT) and other manufacturing exports like textiles. This interdependence is rather high for Indian economy now because we do not have production technologies for high quality, large scale manufacturing for meeting the global standards. Similarly we do not have India-made high end, high value software products or IPR's; hence we buy or get them paying heavy license fees. Technological strengths in industries and institutions are crucial for winning in competition; it is weak in India.

No country in the current phase of human history, when products are large in number, complexities are many and users are diverse, can have self-sufficiency in technology. There will be some imports as machine or human knowledge (consultancies) or IPR's. **But if interdependence is not BALANCED, there is a problem for technology buyer and their businesses.**

In India one of our key deficiency for global competition is TECHNOLOGY in almost all fields of manufacturing, ICT (which is totally dependent on import of equipment and software), infrastructure related technologies (road, rail, shipping), electric power etc. Pharma and auto sector are the ones where the foreign technologies are still dominant but the Indian counter parts have mastered the use of them to produce automobiles mostly for domestic consumption, auto components (for exports as well) and for manufacturing and exporting high quality (compliant to USA and Europe regulations) generic drugs. India is yet to succeed in inventing new drugs/medicines to reach production stages. That creates an excessive dependence on generic manufacture only or licence production,(with low profit margins). Similarly India designed cars are yet to make a domestic mark (Indica, Indigo and Nano are good attempts); but it is foreign car makers who dominate the Indian markets. Even Maruti is now going to be fully dependent on Suzuki to make cars in India for sale to Maruti (It is an outsourcing to a foreign company in the Indian soil by an Indian company after many years of its existence).

There are some local firms like Jain Irrigation who are emerging in the agricultural sector as a big company within India, mostly with their technologies. In the renewable energy sector again dependence on foreign sources for technologies and products is very high. That is the reason why the Government has to give many concessions to the foreign companies to set up shops for "Make in India". Only when foreign companies set up production shops in India, the products can be exported. It is mostly a one-way dependence. Interdependence factor comes in because India is a huge market for many products of the companies. The production shops which are set up in India for Make in India, which are no doubt of benefit to a better degree in creation of jobs in India for Indian UMC's, MC's LMC's and the BB. (Upper middle class, middle class, lower middle class and the Bulky Base). But without internal technological strengths created and sustained, these production shops will have to continue to import technologies and retrain/reskill the workforce (labour). This will of course, benefit the foreign supplier as (s)he has an assured market in India. Thus there is an interdependence. However if such an imbalance continues, in the global competition India's value addition will be in a low or middle income trap. It is evident in the IT sector where the Indian employees are still in the lower end of the value chain, in global comparison. If India can create some global market upgrades to the production lines or innovate new products which would give special competitive edge, then the foreign suppliers of technologies (as machines, tools, know-how etc.,) would have to adopt them as otherwise their competitors will. This is where the global competition helps. **India can significantly tilt the balance of interdependence in its favour (to move away from the low-value chain, low to middle income traps), if it can create/ acquire technologies with its own labour (intellectual and through skilled workforce) in India.**

Even though CAPITAL of Indians is in much better position as there are many high net worth individuals in India, especially after liberalization and they invest also in foreign countries (including developed ones), in the long run the technology-skilled labour-trap would affect their rate of growth. That is the reason why India has not produced a global level mega business. It is through global mega businesses, one can take control of more CAPITAL and have extra clout in global competition.

China Comparison

Look at many Chinese big and mega companies in the global scene, in a wide range of fields. They have, since the mid 1970's intelligently used their huge dependence on foreign companies for technologies and capital, into an advantage. China became the manufacturing base for almost all products of modern economy: steel, electronics, telecom, computers, chemicals, pharma etc. But China converted the heavily loaded interdependence against them (they were in the low end; where there was high – the end product company in China it was fully under control of foreign investors in China) by creating local Chinese companies which have their own technologies and grow as a competitor first in their local markets and then go global. Chinese companies produce a large range of products from domestic white goods to power equipment including nuclear power to trains to computers etc., and have also entered in global e-commerce by the strength of their manufacturing base.

Now as many things in the world CAPITAL or PRODUCTS depend on China, developed world would not like to see China slow down as it will affect their exports. Devaluation of Chinese currency sends shivers in the Capital markets in the developed countries (and thus affects India too). **That is because of the achievement of INTERDEPENDENCE through the systematic methods of modern COMPETITION.** A not-so-visible role for this global competition is also the military strength which has strong components of self-reliance in critical products (nuclear arms, missiles, aircraft, naval carriers, submarines, and missile shields and some pre-emptive attack capabilities as well).

The Chinese story is different from that of the oil-rich Middle-East Countries which accumulated huge CAPITAL as petro dollars (through sale of oil) and lived on it, without acquiring **technological and human resources (skilled labour) strengths within their countries.** They bought modern military equipment from USA and Europe. This interdependence which involved a lot of CAPITAL from Middle-East going to the West and their (middle east countries) total dependence on the West for technology, military and knowledge base of labour, suited USA and Europe very well, to the detriment of the middle-east countries and its people.

Middle East countries are now sitting on fragile economies though a few high net worth individuals from these countries may go and live in Europe, USA etc. ; most of the middle-east countries' people will get into lower middle income trap in the competition with China, perhaps India (if it wakes up and mends its ways) and a few other emerging countries.

Impact on Societies

We have looked at some examples of how the two major elements of modern economy CAPITAL and TECHNOLOGY (skilled and oriented intellectual and work force), shape the economic (and indirectly political) power balances between countries (and even within societies). These two elements are products of COMPETITION and they also shape COMPETITION.

Due to the high complexity of modern technological products (be it a foot wear or a machine or mobile or domestic appliance or biomedical equipment or military hardware) there is a strong INTERDEPENDENCE between owners of technological knowledge and those who can use them effectively for global creation and sale of products or services (i.e., those who operate at different segments of the value chain: an Indian BPO servicing global aviation ticketing services or back office accountancy or exporting textile mills like in Tiruppuretc). Countries like India which are huge in size and population need a balance between various elements of competition and interdependence. If there is a heavy imbalance, one always being at the low end and the other at high end, then interdependence collapses over a period of time. There will be a strong dependence of the low end society on the high end society. Having CAPITAL alone in a low end operating country will not suffice as we saw in the example of middle-east countries.

Also within countries, say for example India, the excessive dependence on IT service sector alone or high end private/ public sector hospitals alone or high end elite educational institutions or financial/real estate sector alone cannot give the competitive edge in knowledge creation and in acquiring global or domestic dominance in a Sector. We should remind ourselves often that the totality of Indian Society has a strong INTERDEPENDENCE. Solving problems of one segment alone will lead to competitive pressures from other segments: as we are noticing for competition for job and educational reservations from different caste, religious groups. Competition for water tends to break the fine balance of interdependence of riparian societies. Actually solutions for this have to be balanced through right **technological applications** for optimal use of water and creating surpluses by continued recycling and reuse.

For avoiding competitive pressures for jobs and admissions, one has to create huge job opportunities with good incomes; for education by liberalising education system to allow **competition** between institutions and streams of study. In a truly competitive market, market forces will help create right interdependence, by having the right comparative advantage to surface. Today it is one-syllabus, one-system of control and accreditation.

Again to project the India's growth rates only in terms of money value CAPITAL, will lead to jobless growth as it had been over the last decade. Even earlier, jobs were not created to match growing and aspirational work force. India has to create **competitive** elements of technologies which do not merely concentrate on increasing productivity but also aim at **job (labour) creating** processes which can still maintain competitive edge in the market. To say it is not possible, is to follow the Western development blindly. They could follow that path as they are dominant for more than three centuries now; they **have the control over global interdependence** thus pushing lower end tasks to developing countries, which may dry up when they succeed with more and more of robotic technologies, 3D printing and such. How they will solve problems within their own societies is yet to be seen. Already there are serious divisive voices against immigration in USA. With so many immigrations from Middle East, Africa and Asia to Europe (which is again a sign of Interdependence of the world, which was "forgotten" when USA and NATO forces with UN "approvals" entered into a bloody war in the middle-east), Europe will find it difficult to retain a fine balance of societal welfare with free market capitalism; to keep to ideals of human rights. And above all the fine balance between competition and interdependence between the 28 European countries which have formed the European Union.

To Watch Further

On the whole many things in the world are churning. Assumptions which were great ideas about half a century ago like freedom for CAPITAL, global free trade which will help free capital flows, the supremacy of technology which will continue to innovate under free market competition and in that process CAPITAL and TECHNOLOGY will create prosperity for societies etc are being shaken, if not yet severely. Earlier to that various socialist models of societal isolationism which believed in curbing capital and technology to the diktats of central ideological commands and the ideas of interdependence controlled by autarkic central command, had failed. China conveniently abandoned them by adopting the idea of market forces, capital led growth competitive technological strengths etc. Still they did not adopt the western concepts of democracy and freedom. They were afraid of the societal divisive forces that may be created in the micro-competition within their societies to gain more "goodies" for themselves. How far they will succeed in the coming decades, is to be seen. Middle-east is in crisis; but the collapse of governance there, would have violent repercussions all over the world – a sure sign of modern day interdependence.

India cannot insulate itself from these forces including various forms of instabilities around its neighbourhood; from the ways in which China tackles her problems. Above all much of the challenges to India are going to be internal as well. The elite and political class appears to prefer the "frog-in-the-well policy" for most of the time but occasionally croaking in the voice of their Western role models, for West driven agenda.

Indians have to discover their situation in the world of competition, interdependence with the global forces of capital, technology, business and military security strategies (which include cyber security, security of space assets) driving them. While India cannot afford to shut her doors any more, it cannot afford to continue with the "mushy-mushy" soft approaches of the past, hoping that time will solve the problems.

The Indian society would have to learn to adapt fast to new situations instead of fighting shadow games in the name of democracy, freedom of speech, "Rights" of various forms. Indian people need better lives which can sustain for a few centuries ahead. Those who care for India and Indian people should wake up and face the challenges of the present to create a better future (near-medium and long term). That would mean jettisoning convenient methods of the past (continuing in the present too).

Still, there are many internal (though dormant) strengths in India. That is my hope in making this appeal to Indians who care.

□□□